T0259716

Ahmed Masmoudi

Design and Electromagnetic Feature Analysis of AC Rotating Machines

 Springer

Ahmed Masmoudi
National School of Engineers of Sfax
University of Sfax
Sfax
Tunisia

ISSN 2191-8112   ISSN 2191-8120   (electronic)
SpringerBriefs in Electrical and Computer Engineering
ISBN 978-981-13-0919-9   ISBN 978-981-13-0920-5   (eBook)
https://doi.org/10.1007/978-981-13-0920-5

Library of Congress Control Number: 2018943720

Printed on acid-free paper

This Springer imprint is published by the registered company Springer Nature Singapore Pte Ltd.
The registered company address is: 152 Beach Road, #21-01/04 Gateway East, Singapore 189721,
Singapore

Preface

Following the 70th oil crisis, the world realized for the first time what it would be like if fuels would no longer be cheap or unavailable. In order to damp the fallouts of such a situation, renewable energies have been the subject of an intensive regain of interest. So many R&D programs were launched so far, with emphasis on the investigation of the power potential of conventional and emergent earth's natural energy reserves.

Moreover, until the 1960s, automotive manufacturers did not worry about the cost of fuel. They had never heard of air pollution and they never thought about the life cycle. Ease of operation with reduced maintenance costs meant everything back then. In recent years, clean air policies are driving the market to embrace new propulsion systems in an attempt to substitute or assist efficiently the internal combustion engine (ICE) by an electric drive unit, yielding respectively the so-called electric and hybrid propulsion systems.

The above sustainable energy and mobility applications consider in most if not all cases a key component that achieves the electromechanical conversion of energy: the electric machine. It operates as a generator which directly converts the wind and wave energies and through a turbine the solar, biomass and geothermal ones, into electricity. It operates as a propeller fed by a battery or a fuel cell pack embedded on board of electric and hybrid vehicles.

This said, it should be underlined that the great penetration of electrical machines in the above-cited and the overwhelming majority of current applications has been made possible thanks to the great progress in the area of machine design. Indeed, during several decades, electrical ac machines have been designed accounting for the fact that they will be connected to the network. This has led to the well-known conventional ac machines (induction and dc-excited synchronous machines) in which the stator windings are sinusoidally distributed in slots around the air gap to couple optimally with the sinusoidal supply.

Starting from the 1980s, the emergence of power electronic converters has removed the need for such a concept as the basis for ac machine design. A new approach based on the principle that the best machine design is the one that simply produces the optimum match between the ac electrical machine and the power electronic converter has led to the so-called "converter-fed machines".

Within this trendy topic, the manuscript is devoted to the design and analysis of the electromagnetic features of ac machines, focusing the involved winding, and covering both grid- and converter-fed machines. The manuscript is structured in three chapters:

- The first one deals with the basis of the design of rotating ac machines with an emphasis on their air gap magnetomotive force (MMF). The survey is initiated by the formulation of the air gap MMF using the Ampere's theorem and the conservation law, followed by the investigation of its harmonic content considering the case of concentrated windings. Then, the case of distributed windings is treated applying the superposition approach that highlights their potentialities in reducing the MMF harmonic content.
- The second chapter is aimed at the formulation of the rotating fields that could be generated considering different techniques involving single and polyphase windings. It also emphasizes the effects of the rotating fields on the winding located on the other side of the air gap. These effects are initiated by the induction of back-EMFs which, at on-load operation, lead to the generation of a second rotating field whose synchronization with the initial one results in the production of an electromagnetic torque.
- The third chapter is dedicated to the design of fractional slot concentrated winding equipping permanent magnet synchronous machines. The study is initiated by the arrangement of the armature winding using the star of slots approach. This latter is then applied for the determination of the winding factors of the back-EMF fundamental and harmonics. A case study is finally treated with emphasis on the armature MMF spatial repartition and harmonic content.

Sfax, Tunisia   Prof. Ahmed Masmoudi
                                                            Head of the Renewable Energies
                                                                and Electric Vehicles Lab.

Contents

About the Author

Ahmed Masmoudi received B.S. from Sfax Engineering National School (SENS), University of Sfax, Sfax, Tunisia, in 1984, Ph.D. from Pierre and Marie Curie University, Paris, France, in 1994, and Research Management Ability from SENS, in 2001, all in electrical engineering. In August 1984, he joined Shlumberger as a field engineer. After this industrial experience, he joined the Tunisian University where he held different positions involved in both education and research activities. He is currently a Professor of Electric Power Engineering at SENS, the Head of the Research Laboratory on Renewable Energies and Electric Vehicles (RELEV) and the Coordinator of the Master on Sustainable Mobility Actuators: Research and Technology. He published up to 85 journal papers, among which 19 were published in IEEE transactions. He presented up 367 papers in international conferences, among which 9 have been presented in plenary sessions, and 3 have been rewarded by the best presented paper prize. He is the co-inventor of a USA patent. He is the Chairman of the Program and Publication Committees of the International Conference on Ecological Vehicles and Renewable Energies (EVER), organized every year in Monte Carlo, Monaco, since 2006. He was also the Chairman of the Technical Program and Publication Committees of the first International Conference on Sustainable Mobility Applications, Renewables, and Technology (SMART) which has been held in Kuwait in November 2015. Its involvement in the above conferences has been marked by an intensive guest-editorship activity with the publication of many special issues of several journals including the IEEE Transactions on Magnetics, COMPEL, ELECTROMOTION, and ETEP. Professor Masmoudi is a Senior Member, IEEE. His main interests include the design of new topologies of ac machines allied to the implementation of advanced and efficient control strategies in drives and generators, applied to renewable energy as well as to electrical automotive systems.

List of Figures

List of Tables

Chapter 1
Air Gap Magnetomotive Force: Formulation and Analysis

Abstract The chapter is aimed at the basis of the design of rotating AC machines with emphasis on their air gap magnetomotive force (MMF). The formulation of the MMF is firstly treated using the Amperes' theorem and the conservation law. Then, the harmonic content of its spatial repartition is investigated. In a second part, different case studies are considered including single and three phase concentrated windings. The third part is devoted to the distributed windings with the superposition approach-based formulation of their MMFs. A special attention is paid to the investigation of the capabilities of the distributed windings in reducing the harmonic content of the MMF.

Keywords Air gap magnetomotive force · Flux conservation law
Superposition theorem · Spatial repartition · *Fourier* expansion
Total harmonic distortion · Concentrated winding · Distributed winding

1.1 Introduction

Up to the beginning of the 1980s, electrical machines have been designed assuming that they will be connected to the grid (grid fed machines). This has led to the well known conventional AC machines, as:

- the squirrel cage induction machine,
- the wound rotor induction machine,
- the DC-excited synchronous machine.

These used to be exclusively connected to the grid through their stator windings which are distributed in slots around the air gap so as to couple optimally with the sinusoidal current supply.

In recent years, the emergence of power electronic converters has removed the need of such a concept as the basis for machine design. Consequently, a new area of electric machine technology has been evolved, based on the principle that the best machine design is that which produces the optimum match between the machine and the power electronic converter, yielding the so-called "converter fed machines".

A. Masmoudi, *Design and Electromagnetic Feature Analysis
of AC Rotating Machines*, SpringerBriefs in Electrical and Computer
Engineering, https://doi.org/10.1007/978-981-13-0920-5_1

Much attention is presently focused on the design of new converter fed machines where many conventional considerations are rethought, such as the number of phases, the number of poles, the winding shape and distribution, the magnetic circuit material and geometry, radial, axial or circumferential flux paths, and so on.

Among the most important design criteria, one can distinguish the arrangement of the machine windings around the air gap. Two major winding arrangements are currently considered:

- the distributed windings in the case where the machine has the same number of pole pairs in both sides of the air gap, leading to a number of slot per pole and per phase equal or higher than unity,
- the fractional-slot concentrated windings in the case where the stator and rotor have different pole pairs, yielding a fractional number of slot per pole and per phase lower than unity.

The present chapter deals with an introduction to AC machine design with emphasis on the winding arrangement. A special attention is paid to the formulation and the analysis of the magnetomotive force (MMF) considering the above-described winding arrangements.

1.2   Formulation and Repartition of the Air Gap MMF Created by Concentrated Windings

1.2.1   Assumptions

For the sake of a simple formulation of the air gap MMF, the following assumptions are commonly considered:

1. the magnetic permeability of the iron μ_{Fe} is supposed to be infinite compared to the air one $\mu_0 = 4\pi 10^{-7}$. Assuming so, the iron could be considered as a magnetic short-circuit with a null MMF across any flux loop portion within the ferromagnetic circuit,
2. the magnetic circuit is supposed linear. Neglecting the magnetic saturation allows the application of linear systems laws especially the superposition theorem,
3. the slot depth is supposed negligible compared to the air gap average radius $\mathcal{R}$. Therefore, the slotting effect could be neglected assuming that the air gap e is constant.

1.2.2   Single Coil MMF Formulation

Let us consider the case of a single coil winding. From a practical point of view, such a winding could equip a DC-excited synchronous generator to achieve the field in the rotor. A cross-section of the machine is shown in Fig. 1.1.

Fig. 1.1 Cross-section for a
synchronous machine
equipped by a single
coil-made field

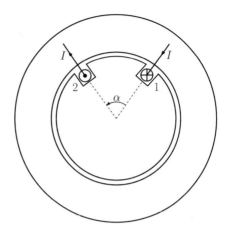

Let us call:

- N the number of turns of the coil,
- I the field current,
- α the slot pitch, that is the angular opening of the coil counted from slot 1 to slot 2 in the anti-clockwise direction.

The MMF $\mathcal{F}$ in a given position θ of the air gap is defined as follows:

$$\mathcal{F} = \int_{air\ gap} Hdl \tag{1.1}$$

with the convention that considers $\mathcal{F}$ positive in the case where the magnetic field $\vec{H}$ is directed from the rotor to the stator and vice-versa.

The formulation of $\mathcal{F}$ goes through the application of:

- the *Ampere* theorem,
- the flux conservation law.

1.2.2.1 Application of the Amperes' Theorem

Accounting for the direction of the circulation of I in the coil of Fig. 1.1, a MMF is created across the magnetic circuit. The resulting flux lines, drawn according to the "tire-bouchon" technic, are shown in Fig. 1.2. One can distinguish two areas in the air gap:

(i) area A where the flux lines are directed from the rotor to the stator,
(ii) area B where the flux lines are directed from the stator to the rotor.

Let us consider the flux line C in Fig. 1.2 which could be subdivided into four parts delimited by the points c_1, c_2, c_3 and c_4 with:

Fig. 1.2 Flux lines through
the magnetic circuit

- c_2 and c_3 correspond to the crossing points from the rotor to the air gap and from the air gap to the stator in the upper part of C, respectively,
- c_4 and c_1 correspond to the crossing points from the stator to the air gap and from the air gap to the rotor in the lower part of C, respectively.

The application of the Amperes' theorem to the flux line C yields:

$$\oint_{(C)} Hdl = NI \tag{1.2}$$

which could be rewritten as follows:

$$\oint_{(C)} Hdl = \int_{c_1}^{c_2} Hdl + \int_{c_2}^{c_3} Hdl + \int_{c_3}^{c_4} Hdl + \int_{c_4}^{c_1} Hdl \tag{1.3}$$

The first assumption of Sect. 1.2.1 leads to:

$$\int_{c_1}^{c_2} Hdl = \int_{c_3}^{c_4} Hdl = 0 \tag{1.4}$$

while

$$\int_{c_2}^{c_3} Hdl = \mathcal{F}_A \tag{1.5}$$

and

$$\int_{c_4}^{c_1} Hdl = -\mathcal{F}_B \tag{1.6}$$

Taking into account Eqs. (1.2), (1.4), (1.5), and (1.6), one can deduce the following:

$$\mathcal{F}_A - \mathcal{F}_B = NI \tag{1.7}$$

1.2.2.2 Application of the Flux Conservation Law

The flux conservation law says that all flux lines crossing the air gap area A return back through the air gap area B, that is to say:

$$\Phi_A + \Phi_B = 0 \tag{1.8}$$

In general, the flux Φ in a given area of the air gap is expressed as:

$$\Phi = SB \tag{1.9}$$

where S is the surface of the considered air gap area and B is its flux density, with:

$$B = \mu_0 H \tag{1.10}$$

Equation (1.1) can be rewritten accounting of the third assumption of Sect. 1.2.1, as follows:

$$\mathcal{F} = eH \tag{1.11}$$

Combining Eqs. (1.10) and (1.11) yields:

$$B = \frac{\mu_0}{e}\mathcal{F} \tag{1.12}$$

Combining Eqs. (1.9) and (1.12) leads to:

$$\Phi = \frac{\mu_0}{e}S\mathcal{F} \tag{1.13}$$

The application of equation (1.13) to areas A and B of the air gap gives the following:

$$\begin{cases} \Phi_A = \frac{\mu_0}{e}S_A\mathcal{F}_A \\ \Phi_B = \frac{\mu_0}{e}S_B\mathcal{F}_B \end{cases} \tag{1.14}$$

where:

$$\begin{cases} S_A = \alpha R L_s \\ S_B = (2\pi - \alpha)R L_s \end{cases} \tag{1.15}$$

where L_s is the machine stuck length.

Equation (1.8) can be rewritten taking into account Eqs. (1.14) and (1.15), as follows:

$$\frac{\mu_0}{e}\alpha R L_s \mathcal{F}_A + \frac{\mu_0}{e}(2\pi - \alpha) R L_s \mathcal{F}_B = 0 \tag{1.16}$$

that leads to:

$$\alpha \mathcal{F}_A + (2\pi - \alpha)\mathcal{F}_B = 0 \tag{1.17}$$

1.2.3 MMF Spatial Repartition

Let us recall Eqs. (1.7) and (1.17):

$$\begin{cases} \mathcal{F}_A \ - \mathcal{F}_B \quad\quad\quad = NI \\ \alpha \mathcal{F}_A + (2\pi - \alpha)\mathcal{F}_B = 0 \end{cases} \tag{1.18}$$

The resolution of such a system leads to:

$$\begin{cases} \mathcal{F}_A = (1 - \frac{\alpha}{2\pi})NI \\ \mathcal{F}_B = -\frac{\alpha}{2\pi}NI \end{cases} \tag{1.19}$$

that gives:

$$\begin{cases} B_A = \frac{\mu_0}{e}(1 - \frac{\alpha}{2\pi})NI \\ B_B = -\frac{\mu_0}{e}\frac{\alpha}{2\pi}NI \end{cases} \tag{1.20}$$

Let us consider the magnetic axis corresponding to the middle of area A, as shown in Fig. 1.2, and let us describe the air gap starting from this axis and turning in the anti-clockwise direction. Doing so along with recording the values of the MMF (respectively the flux density) in each point of the air gap yields the spatial repartition of the MMF (respectively of the flux density).

Figure 1.3 illustrates the spatial repartition of the MMF along the air gap.

The MMF spatial repartition can be redrawn considering a rotation in the same direction but starting from the position $-\pi$ with respect to the magnetic axis, as shown in Fig. 1.4. One can notice that the MMF spatial repartition is an even function of the position θ with a null average value and a angular frequency of the fundamental equal to air gap length (2π). In other words, the fundamental flux density $B_1(\theta)$ is bipolar with a pole angular opening, also called pole pitch and noted τ_p, equal to π.

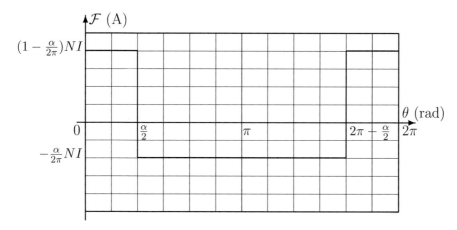

Fig. 1.3 MMF spatial repartition

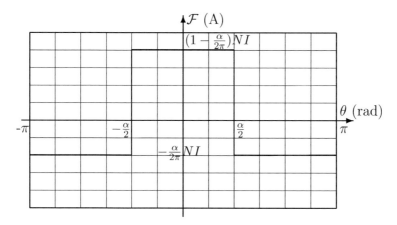

Fig. 1.4 Spatial repartition highlighting the fact that $\mathcal{F}$ is an even function of θ

1.2.3.1 Air Gap MMF *Fourier* Expansion

Let us call β the ratio of the slot pitch α to the pole pitch τ_p which corresponds to the angular opening of a pole of the fundamental FFM, as:

$$\beta = \frac{\alpha}{\tau_p} \tag{1.21}$$

In the case of as single coil winding, the pole pitch $\tau_p = \pi$.

The MMF spatial repartition can be formulated as follows:

$$\mathcal{F}(\theta) = \begin{cases} (1 - \frac{\beta}{2})NI & -\frac{\alpha}{2} \leq \theta \leq \frac{\alpha}{2} \\ -\frac{\beta}{2}NI & -\pi \leq \theta \leq -\frac{\alpha}{2} \ \& \ \frac{\alpha}{2} \leq \theta \leq \pi \end{cases} \tag{1.22}$$

The general term F_n of its *Fourier* transform is calculated as follows:

$$F_n = \frac{2}{\pi} \left(\int_0^{\frac{\pi}{2}\beta} (1 - \frac{\beta}{2})NI \cos n\theta d\theta + \int_{\frac{\pi}{2}\beta}^{\pi} -\frac{\beta}{2}NI \cos n\theta d\theta \right) \tag{1.23}$$

which leads to:

$$F_n = \frac{2}{\pi} NI \frac{\sin n\frac{\pi}{2}\beta}{n} \tag{1.24}$$

and finally:

$$\mathcal{F}(\theta) = \sum_{n=1}^{n=\infty} \frac{2}{\pi} NI \frac{\sin n\frac{\pi}{2}\beta}{n} \cos n\theta \tag{1.25}$$

In order to characterize the harmonic content of $\mathcal{F}(\theta)$, one can consider its total harmonic distortion *THD*, such as:

$$THD(\%) = 100 \frac{\sqrt{\sum_{n=2}^{n=\infty} F_n^2}}{F_1} = 100 \frac{\sqrt{\sum_{n=2}^{n=\infty} \left(\frac{\sin n\frac{\pi}{2}\beta}{n} \right)^2}}{\sin \frac{\pi}{2}\beta} \tag{1.26}$$

Fig. 1.5 shows the variation of the *THD* with respect to β.

1.2.3.2   Case of a Diametral Coil

Referring to Fig. 1.5, it clearly appears that a unity slot pitch to the pole pitch ratio β offers the lowest value of the THD of the air gap MMF. This case is characterized by $\alpha = \pi$, yielding the so-called "diametral coil".

The formulation of the MMF spatial repartition turns to be as follows:

$$\mathcal{F}(\theta) = \begin{cases} \frac{NI}{2} & -\frac{\pi}{2} \leq \theta \leq \frac{\pi}{2} \\ -\frac{NI}{2} & -\pi \leq \theta \leq -\frac{\pi}{2} \ \& \ \frac{\pi}{2} \leq \theta \leq \pi \end{cases} \tag{1.27}$$

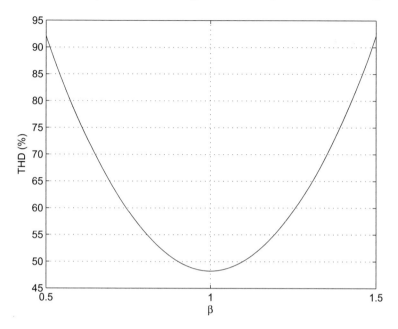

Fig. 1.5 THD of the MMF spatial repartition versus β

The general term F_n of its *Fourier* expansion is expressed as follows:

$$F_n = \frac{2}{\pi} NI \frac{\sin n\frac{\pi}{2}}{n} = \begin{cases} \frac{2}{\pi} NI \frac{1}{n} & \text{for } n = 1, 5, 9, \ldots \\ -\frac{2}{\pi} NI \frac{1}{n} & \text{for } n = 3, 7, 11, \ldots \\ 0 & \text{for } n = 2, 4, 6, \ldots \end{cases} \quad (1.28)$$

which could be rewritten as:

$$F_{2n+1} = \frac{2}{\pi} NI \frac{(-1)^n}{(2n+1)} \quad \text{for } n = 0, 1, 2, 3, 4, \ldots \quad (1.29)$$

and then the *Fourier* enpension is expressed as:

$$\mathcal{F}(\theta) = \sum_{n=0}^{n=\infty} \frac{2}{\pi} NI \frac{(-1)^n}{(2n+1)} \cos(2n+1)\theta \quad (1.30)$$

Although the diametral winding exhibits the lowest harmonic pollution, its *THD* remains high equal to 48.3167% with a remarkable dominance of the harmonics of low ranks. Indeed, the *Fourier* expansion of $\mathcal{F}(\theta)$ could be rewritten as:

$$\mathcal{F}(\theta) = \frac{2}{\pi} NI \left(\cos\theta - \frac{1}{3}\cos 3\theta + \frac{1}{5}\cos 5\theta - \frac{1}{7}\cos 7\theta + \ldots + \frac{(-1)^n}{(2n+1)}\cos(2n+1)\theta \right) \quad (1.31)$$

Fig. 1.6 Cross-section of a
synchronous machine
equipped by a dual
coil-made field

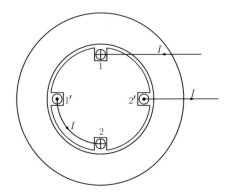

that gives the following normalized amplitudes of the harmonics of low ranks:

$$\begin{cases} \left|\dfrac{F_3}{F_1}\right| = \dfrac{1}{3} \\[2mm] \left|\dfrac{F_5}{F_1}\right| = \dfrac{1}{5} \\[2mm] \left|\dfrac{F_7}{F_1}\right| = \dfrac{1}{7} \end{cases} \tag{1.32}$$

1.2.4   Case of a Dual Diametral Coil Winding

This case is illustrated in Fig. 1.6 where the field of the synchronous machine is made
up of two coils connected in series with N turns each inserted in two equidistant pair
of slots 1-1' and 2-2'. Such a winding is characterized by:

$$\begin{cases} \alpha = \dfrac{\pi}{2} \\ \beta = 1 \end{cases} \tag{1.33}$$

Referring to Fig. 1.6, one can distinguish four areas in the air gap: A, B, C, and
D. In order to formulate the air gap MMF, the superposition approach is applied
assuming, in a first step, the case where just the coil inserted in the couple of slots
1-1' is supplied by the field current I.

The application of the Amperes' theorem gives the following:

$$\begin{cases} \mathcal{F}_A - \mathcal{F}_D = NI & \text{considering the flux line } C_1 \\ \mathcal{F}_A - \mathcal{F}_B = NI & \text{considering the flux line } C_2 \\ \mathcal{F}_C - \mathcal{F}_B = 0 & \text{considering the flux line } C_3 \\ \mathcal{F}_C - \mathcal{F}_D = 0 & \text{considering the flux line } C_4 \end{cases} \tag{1.34}$$

The application of the flux conservation law yields:

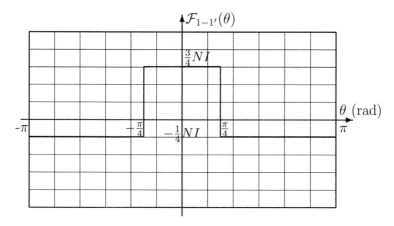

Fig. 1.7 Spatial repartition of the MMF $\mathcal{F}_{1-1'}$ produced by the coil in the slot pair 1-1'

$$\mathcal{F}_A + \mathcal{F}_B + \mathcal{F}_C + \mathcal{F}_D = 0 \tag{1.35}$$

The resolution of equations (1.34) and (1.35) has led to:

$$\begin{cases} \mathcal{F}_A = \frac{3}{4}NI \\ \mathcal{F}_B = -\frac{1}{4}NI \\ \mathcal{F}_C = -\frac{1}{4}NI \\ \mathcal{F}_D = -\frac{1}{4}NI \end{cases} \tag{1.36}$$

This solution could be found in another manner considering the formulation treated in Sect. 1.2.2, with:

$$\begin{cases} \alpha = \frac{\pi}{2} \\ \beta = \frac{1}{2} \end{cases} \tag{1.37}$$

and with areas B, C, and D combined in a single area called B.

Considering the magnetic axis of the coil inserted in 1-1' as the origin of angles θ, the application of the equation (1.22) gives:

$$\mathcal{F}_{1-1'}(\theta) = \begin{cases} \frac{3}{4}NI & -\frac{\pi}{4} \leq \theta \leq \frac{\pi}{4} \\ -\frac{1}{4}NI & -\pi \leq \theta \leq -\frac{\pi}{4} \ \& \ \frac{\pi}{4} \leq \theta \leq \pi \end{cases} \tag{1.38}$$

which confirms the expressions given in Eq. (1.34).

The resulting spatial repartition is shown in Fig. 1.7.

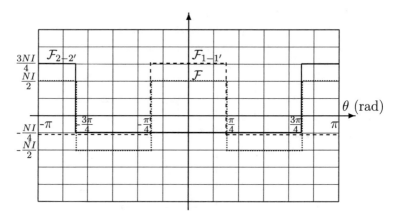

Fig. 1.8 Spatial repartition of the MMF $\mathcal{F}$ obtained following the superposition of $\mathcal{F}_{1-1'}$ and $\mathcal{F}_{2-2'}$. **Legend:** (continuous line) $\mathcal{F}_{2-2'}$, (interrupted line) $\mathcal{F}_{1-1'}$, (dashed line) $\mathcal{F}$

Now, let us assume the case where just the coil inserted in the couple of slots 2-2' is supplied by the field current I. In this case, the area C surrounded by the supplied coil behaves like area A in the previous case, producing an MMF of $\frac{3}{4}NI$. Thus, one can deduce the following:

$$\begin{cases} \mathcal{F}_A = -\frac{1}{4}NI \\ \mathcal{F}_B = -\frac{1}{4}NI \\ \mathcal{F}_C = \frac{3}{4}NI \\ \mathcal{F}_D = -\frac{1}{4}NI \end{cases} \tag{1.39}$$

Considering the magnetic axis of the coil inserted in 1-1' as the origin of angles θ, the resulting MMF spatial repartition is expressed as follows:

$$\mathcal{F}_{2-2'}(\theta) = \begin{cases} \frac{3}{4}NI & -\pi \le \theta \le -\frac{3\pi}{4} \ \& \ \frac{3\pi}{4} \le \theta \le \pi \\ -\frac{1}{4}NI & -\frac{3\pi}{4} \le \theta \le \frac{3\pi}{4} \end{cases} \tag{1.40}$$

The spatial repartition representation of the MMF $\mathcal{F}_{2-2'}$ is illustrated in Fig. 1.8; while the one of the MMF $\mathcal{F}_{1-1'}$ is recalled in interrupted lines. The resultant MMF $\mathcal{F}$ is shown in dashed mixed lines in the same figure.

One can notice that $\mathcal{F}(\theta)$ presents two periods along the air gap with a pole pitch $\tau_p = \frac{\pi}{2}$. Similarly, the flux density is characterized by a two pole pairs. In order to characterize the difference between the mechanical periodicity related to the coil repartition and the electrical one related to the MMF and flux density spatial

Fig. 1.9 Cross-section of a synchronous machine equipped by a multi coil-made field

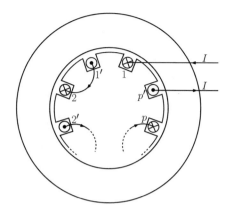

repartitions, the so-called "electrical angle", noted θ_e, is introduced, such that:

$$\theta_e = 2\theta \tag{1.41}$$

The *Fourier* expansion is expressed in terms of the electrical angle as:

$$\mathcal{F}(\theta_e) = \sum_{n=0}^{n=\infty} \frac{2}{\pi} NI \frac{(-1)^n}{(2n+1)} \cos(2n+1)\theta_e \tag{1.42}$$

or in terms of the mechanical one as:

$$\mathcal{F}(\theta) = \sum_{n=0}^{n=\infty} \frac{2}{\pi} NI \frac{(-1)^n}{(2n+1)} \cos(2n+1)2\theta \tag{1.43}$$

1.2.5   Case of a Multi Diametral Coil Winding

This case is illustrated in Fig. 1.9 where the field of the synchronous machine is made up of p coils connected in series with N turns each inserted in p equidistant pairs of slots 1-1', 2-2', ..., p-p'. Such a winding is characterized by:

$$\begin{cases} \alpha = \dfrac{\pi}{p} \\ \beta = 1 \end{cases} \tag{1.44}$$

The MMF spatial repartition, resulting from feeding such a winding, could be deduced from the ones obtained in the cases of a single and double coils, as shown in Fig. 1.10.

Referring to Fig. 1.10, one can notice that the fundamental MMF presents p periods, that is to say that the flux density exhibits p pole pairs, with a pole pitch $\tau_p = \frac{\pi}{p}$.

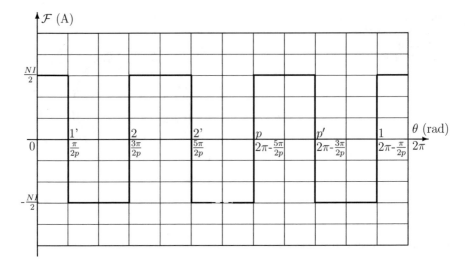

Fig. 1.10 Spatial repartition of the MMF produced by the field of the machine of Fig. 1.9

Consequently, the electrical angle θ_e is expressed in terms of the mechanical one θ as:

$$\theta_e = p\theta \tag{1.45}$$

The *Fourier* expansion is expressed in terms of the electrical angle as:

$$\mathcal{F}(\theta_e) = \sum_{n=0}^{n=\infty} \frac{2}{\pi} NI \frac{(-1)^n}{(2n+1)} \cos(2n+1)\theta_e \tag{1.46}$$

or in terms of the mechanical one as:

$$\mathcal{F}(\theta) = \sum_{n=0}^{n=\infty} \frac{2}{\pi} NI \frac{(-1)^n}{(2n+1)} \cos(2n+1)p\theta \tag{1.47}$$

1.2.6 Three Phase Windings

A three phase winding is a set of three identical circuits, called phases and noted "*a*", "*b*" and "*c*", which are arranged within the air gap in slots in such away that their respective positions are shifted by a 120°-electrical angle. For the sake of simplicity, the cases treated in this paragraph concern three phase windings where each phase is made up of coils of N turns concentrated in a single pair of slots.

Fig. 1.11 Cross-section of a machine equipped by a three phase winding with $\alpha < \frac{2\pi}{3}$

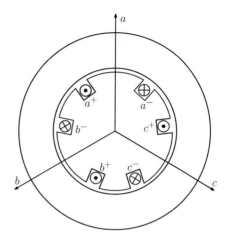

1.2.6.1 Case of One Coil per Phase

MMF Spatial Repartition Let consider the case where $\alpha < \frac{2\pi}{3}$ in such a way that the resulting slot occupation is: $a^+ - b^- - b^+ - c^- - c^+ - a^-$, as illustrated in Fig. 1.11.

Under a sinusoidal current supply such as:

$$\begin{cases} i_a = \sqrt{2}I_{rms}\cos\omega t \\ i_b = \sqrt{2}I_{rms}\cos(\omega t - \frac{2\pi}{3}) \\ i_c = \sqrt{2}I_{rms}\cos(\omega t + \frac{2\pi}{3}) \end{cases} \tag{1.48}$$

One can derive and represent the spatial repartition of the air gap MMF for a given instant. Let us consider for instance the initial time ($t = 0$), the phase currents turn to be:

$$\begin{cases} i_a = \sqrt{2}I_{rms} \qquad\quad = I_{max} \\ i_b = \sqrt{2}I_{rms}(-\frac{1}{2}) = -\frac{1}{2}I_{max} \\ i_c = \sqrt{2}I_{rms}(-\frac{1}{2}) = -\frac{1}{2}I_{max} \end{cases} \tag{1.49}$$

The MMF spatial repartition created by the "a" phase is expressed as follows:

$$\mathcal{F}_a(\theta) = \begin{cases} (1 - \frac{\beta}{2})NI_{max} & -\frac{\alpha}{2} \leq \theta \leq \frac{\alpha}{2} \\ -\frac{\beta}{2}NI_{max} & -\pi \leq \theta \leq -\frac{\alpha}{2} \;\&\; \frac{\alpha}{2} \leq \theta \leq \pi \end{cases} \tag{1.50}$$

The MMF spatial repartition created by the "b" phase is expressed as follows:

$$\mathcal{F}_b(\theta) = \begin{cases} -\frac{1}{2}(1 - \frac{\beta}{2})NI_{max} & -\frac{\alpha}{2} + \frac{2\pi}{3} \le \theta \le \frac{\alpha}{2} + \frac{2\pi}{3} \\ \frac{1}{2}\frac{\beta}{2}NI_{max} & -\pi \le \theta \le -\frac{\alpha}{2} + \frac{2\pi}{3} \ \ \& \ \frac{\alpha}{2} + \frac{2\pi}{3} \le \theta \le \pi \end{cases} \tag{1.51}$$

The MMF spatial repartition created by the "c" phase is expressed as follows:

$$\mathcal{F}_c(\theta) = \begin{cases} -\frac{1}{2}(1 - \frac{\beta}{2})NI_{max} & -\frac{\alpha}{2} - \frac{2\pi}{3} \le \theta \le \frac{\alpha}{2} - \frac{2\pi}{3} \\ \frac{1}{2}\frac{\beta}{2}NI_{max} & -\pi \le \theta \le -\frac{\alpha}{2} - \frac{2\pi}{3} \ \ \& \ \frac{\alpha}{2} - \frac{2\pi}{3} \le \theta \le \pi \end{cases} \tag{1.52}$$

The resultant MMF spatial repartition is obtained by summing the ones created by the three phases, such that:

$$\mathcal{F}(\theta) = \begin{cases} 0 & -\pi \le \theta \le -\frac{\alpha}{2} - \frac{2\pi}{3} \\ -\frac{1}{2}NI_{max} & -\frac{\alpha}{2} - \frac{2\pi}{3} \le \theta \le \frac{\alpha}{2} - \frac{2\pi}{3} \\ 0 & \frac{\alpha}{2} - \frac{2\pi}{3} \le \theta \le -\frac{\alpha}{2} \\ NI_{max} & -\frac{\alpha}{2} \le \theta \le \frac{\alpha}{2} \\ 0 & \frac{\alpha}{2} \le \theta \le -\frac{\alpha}{2} + \frac{2\pi}{3} \\ -\frac{1}{2}NI_{max} & -\frac{\alpha}{2} + \frac{2\pi}{3} \le \theta \le \frac{\alpha}{2} + \frac{2\pi}{3} \\ 0 & \frac{\alpha}{2} + \frac{2\pi}{3} \le \theta \le \pi \end{cases} \tag{1.53}$$

MMF *Fourier* Expansion The *Fourier* expansions of the MMFs created by the three phases are expressed as:

$$\begin{cases} \mathcal{F}_a(\theta) = \sum_{n=1}^{n=\infty} \frac{2}{\pi}NI_{max}\frac{\sin n\frac{\pi}{2}\beta}{n} \cos n\theta \\ \mathcal{F}_b(\theta) = \sum_{n=1}^{n=\infty} \frac{2}{\pi}(-\frac{1}{2})NI_{max}\frac{\sin n\frac{\pi}{2}\beta}{n} \cos(n(\theta - \frac{2\pi}{3})) \\ \mathcal{F}_c(\theta) = \sum_{n=1}^{n=\infty} \frac{2}{\pi}(-\frac{1}{2})NI_{max}\frac{\sin n\frac{\pi}{2}\beta}{n} \cos(n(\theta + \frac{2\pi}{3})) \end{cases} \tag{1.54}$$

The superposition yields the *Fourier* expansion of the resultant MMF $\mathcal{F}(\theta)$, as follows:

$$\mathcal{F}(\theta) = \sum_{n=1}^{n=\infty} \frac{2}{\pi}NI_{max}\frac{\sin n\frac{\pi}{2}\beta}{n}\left(1 - \cos(n\frac{2\pi}{3})\right) \cos n\theta \tag{1.55}$$

One can notice the general amplitude F_n of $\mathcal{F}(\theta)$ takes two values, such that:

Fig. 1.12 Cross-section of a machine equipped by a three phase winding with $\alpha = \pi$

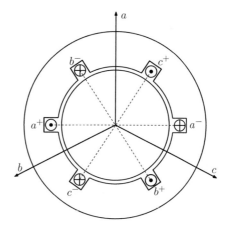

$$F_n = \begin{cases} 0 & \text{for } n \text{ multiple of 3} \\ \dfrac{2}{\pi}\dfrac{3}{2}NI_{max}\dfrac{\sin n\frac{\pi}{2}\beta}{n} & \text{for } n \text{ non multiple of 3} \end{cases} \qquad (1.56)$$

Case of Diametral Coils This case is characterized by the following relations:

$$\begin{cases} \alpha = \pi \\ \beta = 1 \end{cases} \qquad (1.57)$$

The resulting slot occupation is: $a^+ - c^- - b^+ - a^- - c^+ - b^-$, as illustrated in Fig. 1.12.

Considering a sinusoidal supply with the maximum current in phase "a", the MMF spatial repartitions of the three phases and their resultant are drawn in Fig. 1.13. One can notice that $\mathcal{F}$ presents two poles, that is to say a pole pair $p = 1$.

The *Fourier* expansion is expressed as follows:

$$\mathcal{F}(\theta) = \sum_{n=1}^{n=\infty} \frac{2}{\pi}NI_{max}\frac{\sin n\frac{\pi}{2}}{n}\left(1 - \cos(n\frac{2\pi}{3})\right)\cos n\theta = \sum_{n=1}^{n=\infty} F_n \cos n\theta \qquad (1.58)$$

where F_n takes the following values:

$$F_n = \begin{cases} 0 & \text{for } n \text{ multiple of 3 and 2} \\ \dfrac{2}{\pi}\dfrac{3}{2}NI_{max}\left(\dfrac{1}{n}\right) & \text{for } n = 1, 5, 13, 17, 25, 29, \ldots \\ \dfrac{2}{\pi}\dfrac{3}{2}NI_{max}\left(-\dfrac{1}{n}\right) & \text{for } n = 7, 11, 19, 23, 31, 35, \ldots \end{cases} \qquad (1.59)$$

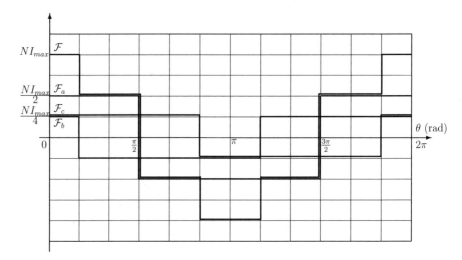

Fig. 1.13 Spatial repartitions of the phase MMFs and their resultant for $p = 1$

1.2.6.2   Case of Two Diametral Coils per Phase

In this case, each phase is made up of two coils of N turns concentrated in a single pair of slots each. The two coils are connected in series in such a way that the winding produces four poles (tetrapolar with the pole pair $p = 2$). This gives $\theta_e = 2\theta$, with:

$$\begin{cases} \alpha = \dfrac{\pi}{2} \\ \tau_p = \dfrac{\pi}{2} \\ \beta = 1 \end{cases} \tag{1.60}$$

Consequently, the $\frac{2\pi}{3}$-electrical shift between the three phases corresponds to a mechanical angle of $\frac{\pi}{3}$. That is to say:

- the slot b_1^+ is shifted from the slot a_1^+ by $\frac{\pi}{3}$ and the slot c_1^+ is shifted from the slot b_1^+ by $\frac{\pi}{3}$,
- the slot b_2^+ is shifted from the slot a_2^+ by $\frac{\pi}{3}$ and the slot c_2^+ is shifted from the slot b_2^+ by $\frac{\pi}{3}$.

The resulting slot occupation is:

$$a_1^+ - c_1^- - b_1^+ - a_2^- - c_1^+ - b_2^- - a_2^+ - c_2^- - b_2^+ - a_1^- - c_2^+ - b_1^-$$

as illustrated in Fig. 1.14.

Fig. 1.14 Cross-section of a machine equipped by a three phase winding with $\alpha = \frac{\pi}{2}$

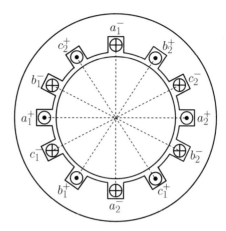

Considering a sinusoidal supply with the maximum current in phase "a", the MMF spatial repartitions of the three phases and their resultant are tetrapolar ($p = 2$) and are a duplication of the ones shown in Fig. 1.13.

The *Fourier* expansions of the MMFs created by the three phases are expressed as:

$$
\begin{cases}
\mathcal{F}_a(\theta_e) = \displaystyle\sum_{n=1}^{n=\infty} \frac{2}{\pi} NI_{max} \frac{\sin n\frac{\pi}{2}}{n} \cos n\theta_e \\[2mm]
\mathcal{F}_b(\theta_e) = \displaystyle\sum_{n=1}^{n=\infty} \frac{2}{\pi}(-\frac{1}{2})NI_{max} \frac{\sin n\frac{\pi}{2}}{n} \cos(n(\theta_e - \frac{2\pi}{3})) \\[2mm]
\mathcal{F}_c(\theta_e) = \displaystyle\sum_{n=1}^{n=\infty} \frac{2}{\pi}(-\frac{1}{2})NI_{max} \frac{\sin n\frac{\pi}{2}}{n} \cos(n(\theta_e + \frac{2\pi}{3}))
\end{cases}
\tag{1.61}
$$

The superposition yields the *Fourier* expansion of the resultant MMF $\mathcal{F}$ which could be expressed in terms of θ_e, as follows:

$$
\mathcal{F}(\theta_e) = \sum_{n=1}^{n=\infty} \frac{2}{\pi} NI_{max} \frac{\sin n\frac{\pi}{2}}{n} \left(1 - \cos(n\frac{2\pi}{3})\right) \cos n\theta_e = \sum_{n=1}^{n=\infty} F_n \cos n\theta_e
\tag{1.62}
$$

or in terms of θ, as follows:

$$
\mathcal{F}(\theta) = \sum_{n=1}^{n=\infty} F_n \cos 2n\theta
\tag{1.63}
$$

where F_n takes the following values:

$$
F_n =
\begin{cases}
0 & \text{for } n \text{ multiple of 3 and 2} \\[2mm]
\frac{2}{\pi}\frac{3}{2}NI_{max}\left(\frac{1}{n}\right) & \text{for } n = 1, 5, 13, 17, 25, 29, \dots \\[2mm]
\frac{2}{\pi}\frac{3}{2}NI_{max}\left(-\frac{1}{n}\right) & \text{for } n = 7, 11, 19, 23, 31, 35, \dots
\end{cases}
\tag{1.64}
$$

1.3 Formulation and Repartition of the Air Gap MMF Created by Distributed Windings

1.3.1 Motivations in Using Distributed Windings

Referring to the case studies previously-treated, one can clearly notice that the harmonic content of the MMF spatial repartitions are far from being sinusoidal. This represents a major drawback for grid-connected machines. Indeed, beyond the fundamental, the harmonics cause the following shortcomings:

- the increase of the machine iron losses,
- the injection of harmonics to the grid that greatly affect the operation of the loads.

For sure, three phase windings have the capability to eradicate the harmonics of ranks multiple of three in general, and those multiple of two in the case where $\beta = 1$. That is to say, only the harmonics of ranks 5, 7, 11, 13, 17, 19, 23, 25, 29, 31 .... remain in the MMF spectrum. This is insufficient especially the harmonic of rank 5 which has an amplitude equal to 20% of the fundamental one. Moreover, in the case of a single phase winding, the harmonic content is higher with a *THD* of the MMF reaching up to 48.316% in the case where $\beta = 1$.

In order to reduce the MMF harmonic content, a solution consisting in distributing each coil, in a number of pair slots higher than unity, has been commonly adopted. Beyond the improvement of the waveform of the MMF, the winding distribution leads to a reduction of the slot size which represents a crucial design benefit in so far as the hypothesis regarding the omission of the slotting effect turns to be more and more satisfied. Moreover, this yields a reduction of the slot leakage flux and consequently an increase of the efficiency. However, increasing the number of slot pairs may be compromised by some limitations, such as:

- a saturation of the teeth,
- a weakness of the mechanical structure which is critical in the case where the distributed winding is linked to the rotor,
- a complexity of the manufacturing.

Different filling techniques of the slots could be adopted. One can distinguish:

- the balanced distribution,
- the linear distribution,
- the sinusoidal distribution.

1.3.2 Case Study

Let us consider the case where the single coil-made field of Fig. 1.1 is distributed in six slot pairs: 1-1', 2-2', 3-3', 4-4', 5-5', and 6-6', with an slot pitch equal to π. The angular shift between two adjacent slots is named γ.

1.3.2.1 Balanced Distribution

Let us consider the case where the six slot pairs are filled uniformly, yielding the so-called "balanced distribution". That is to say that the N turns of the coil are equally-distributed on the six slot pairs leading to $\frac{N}{6}$ in each one, as illustrated in Fig. 1.15.

Feeding the distributed winding by the DC current I, the spatial repartitions of the MMFs produced by the coils located in the six slot pairs are deduced from the one of Fig. 3.19 for $\alpha = \pi$. The obtained MMFs as well as their resultant $\mathcal{F}_B$ are shown in Fig. 1.16.

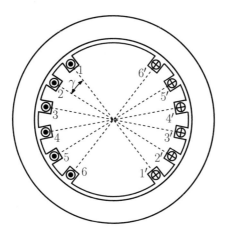

Fig. 1.15 Cross-section of a synchronous machine equipped by a single coil-made field equally-distributed in six slot pairs with $\beta = 1$

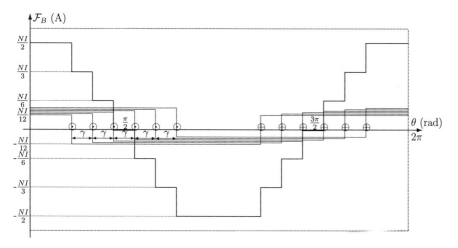

Fig. 1.16 Spatial repartitions of the MMFs created by the coils and their resultant $\mathcal{F}_B$

The resulting MMF spatial repartitions could be expressed as follows:

$$
\begin{cases}
\dfrac{NI}{12}
\begin{cases}
0 \le \theta \le \dfrac{(\pi - 5\gamma)}{2} & \& \quad \dfrac{(3\pi - 5\gamma)}{2} \le \theta \le 2\pi \text{ for } \mathcal{F}_{1-1'} \\[2mm]
0 \le \theta \le \dfrac{(\pi - 3\gamma)}{2} & \& \quad \dfrac{(3\pi - 3\gamma)}{2} \le \theta \le 2\pi \text{ for } \mathcal{F}_{2-2'} \\[2mm]
0 \le \theta \le \dfrac{(\pi - \gamma)}{2} & \& \quad \dfrac{(3\pi - \gamma)}{2} \le \theta \le 2\pi \text{ for } \mathcal{F}_{3-3'} \\[2mm]
0 \le \theta \le \dfrac{(\pi + \gamma)}{2} & \& \quad \dfrac{(3\pi + \gamma)}{2} \le \theta \le 2\pi \text{ for } \mathcal{F}_{4-4'} \\[2mm]
0 \le \theta \le \dfrac{(\pi + 3\gamma)}{2} & \& \quad \dfrac{(3\pi + 3\gamma)}{2} \le \theta \le 2\pi \text{ for } \mathcal{F}_{5-5'} \\[2mm]
0 \le \theta \le \dfrac{(\pi + 5\gamma)}{2} & \& \quad \dfrac{(3\pi + 5\gamma)}{2} \le \theta \le 2\pi \text{ for } \mathcal{F}_{6-6'}
\end{cases} \\[20mm]
-\dfrac{NI}{12}
\begin{cases}
\dfrac{(\pi - 5\gamma)}{2} \le \theta \le \dfrac{3\pi - 5\gamma}{2} \text{ for } \mathcal{F}_{1-1'} \\[2mm]
\dfrac{(\pi - 3\gamma)}{2} \le \theta \le \dfrac{3\pi - 3\gamma}{2} \text{ for } \mathcal{F}_{2-2'} \\[2mm]
\dfrac{(\pi - \gamma)}{2} \le \theta \le \dfrac{3\pi - \gamma}{2} \text{ for } \mathcal{F}_{3-3'} \\[2mm]
\dfrac{(\pi + \gamma)}{2} \le \theta \le \dfrac{3\pi + \gamma}{2} \text{ for } \mathcal{F}_{4-4'} \\[2mm]
\dfrac{(\pi + 3\gamma)}{2} \le \theta \le \dfrac{3\pi + 3\gamma}{2} \text{ for } \mathcal{F}_{5-5'} \\[2mm]
\dfrac{(\pi + 5\gamma)}{2} \le \theta \le \dfrac{3\pi + 5\gamma}{2} \text{ for } \mathcal{F}_{6-6'}
\end{cases}
\end{cases}
\tag{1.65}
$$

Their *Fourier* expansions are derived taking into account their angular shifts with respect to the reference positions $\frac{\pi}{2}$ and $\frac{3\pi}{2}$, as follows:

$$
\begin{cases}
\mathcal{F}_{1-1'}(\theta) = \displaystyle\sum_{n=0}^{n=\infty} \frac{2}{\pi}\frac{NI}{6}\frac{(-1)^n}{(2n+1)} \cos(2n+1)(\theta + 5\tfrac{\gamma}{2}) \\[4mm]
\mathcal{F}_{2-2'}(\theta) = \displaystyle\sum_{n=0}^{n=\infty} \frac{2}{\pi}\frac{NI}{6}\frac{(-1)^n}{(2n+1)} \cos(2n+1)(\theta + 3\tfrac{\gamma}{2}) \\[4mm]
\mathcal{F}_{3-3'}(\theta) = \displaystyle\sum_{n=0}^{n=\infty} \frac{2}{\pi}\frac{NI}{6}\frac{(-1)^n}{(2n+1)} \cos(2n+1)(\theta + \tfrac{\gamma}{2}) \\[4mm]
\mathcal{F}_{4-4'}(\theta) = \displaystyle\sum_{n=0}^{n=\infty} \frac{2}{\pi}\frac{NI}{6}\frac{(-1)^n}{(2n+1)} \cos(2n+1)(\theta - \tfrac{\gamma}{2}) \\[4mm]
\mathcal{F}_{5-5'}(\theta) = \displaystyle\sum_{n=0}^{n=\infty} \frac{2}{\pi}\frac{NI}{6}\frac{(-1)^n}{(2n+1)} \cos(2n+1)(\theta - 3\tfrac{\gamma}{2}) \\[4mm]
\mathcal{F}_{6-6'}(\theta) = \displaystyle\sum_{n=0}^{n=\infty} \frac{2}{\pi}\frac{NI}{6}\frac{(-1)^n}{(2n+1)} \cos(2n+1)(\theta - 5\tfrac{\gamma}{2})
\end{cases}
\tag{1.66}
$$

which could be partially summed as:

$$
\begin{cases}
\mathcal{F}_{1-1'}(\theta) + \mathcal{F}_{6-6'}(\theta) = \sum_{n=0}^{n=\infty} \frac{2}{\pi} \frac{NI}{3} \frac{(-1)^n}{(2n+1)} \cos(2n+1)5\frac{\gamma}{2} \cos(2n+1)\theta \\[2mm]
\mathcal{F}_{2-2'}(\theta) + \mathcal{F}_{5-5'}(\theta) = \sum_{n=0}^{n=\infty} \frac{2}{\pi} \frac{NI}{3} \frac{(-1)^n}{(2n+1)} \cos(2n+1)3\frac{\gamma}{2} \cos(2n+1)\theta \quad (1.67)\\[2mm]
\mathcal{F}_{3-3'}(\theta) + \mathcal{F}_{4-4'}(\theta) = \sum_{n=0}^{n=\infty} \frac{2}{\pi} \frac{NI}{3} \frac{(-1)^n}{(2n+1)} \cos(2n+1)\frac{\gamma}{2} \cos(2n+1)\theta
\end{cases}
$$

Finally, the total sum leads to the *Fourier* expansion of the resultant MMF $\mathcal{F}_B(\theta)$ as:

$$
\mathcal{F}_B(\theta) = \sum_{n=0}^{n=\infty} \frac{2}{\pi} \frac{NI}{3} \frac{(-1)^n}{(2n+1)} \left(\cos(2n+1)5\frac{\gamma}{2} + \cos(2n+1)3\frac{\gamma}{2} + \cos(2n+1)\frac{\gamma}{2} \right) \cos(2n+1)\theta \quad (1.68)
$$

The benefit of the distribution on the quality of the MMF is highlighted by selecting the angular shift γ that enables the cancelation of the most crucial harmonic; the closest to the fundamental. In the treated case, the concerned harmonic in the one of rank 3. The relation allowing its cancelation is obtained from equation (1.68) for $n = 1$, leading to:

$$
\cos 15\frac{\gamma}{2} + \cos 9\frac{\gamma}{2} + \cos 3\frac{\gamma}{2} = 0 \quad (1.69)
$$

Its resolution gives $\gamma = \frac{\pi}{9} = 20°$.

In order to check the appropriateness of such a solution, let us consider the variation of the total harmonic distortion THD_B with respect to γ, as:

$$
THD_B(\%) = 100 \frac{\sqrt{\sum_{n=1}^{n=\infty} \left(\dfrac{\cos(2n+1)5\frac{\gamma}{2} + \cos(2n+1)3\frac{\gamma}{2} + \cos(2n+1)\frac{\gamma}{2}}{2n+1} \right)^2}}{\cos 5\frac{\gamma}{2} + \cos 3\frac{\gamma}{2} + \cos \frac{\gamma}{2}} \quad (1.70)
$$

The obtained results are shown in Fig. 1.17. One can notice $\gamma = 20°$ is almost the angle corresponding to the minimum of THD_B, with $\min(THD_B) \simeq 11.77\%$.

1.3.2.2   Linear Distribution

Let us consider the case where the six slot pairs are filled in such a way that the number of turns is increased with the same ratio in the slot pairs located before the reference positions $\frac{\pi}{2}$ and $\frac{3\pi}{2}$, and inversely in the slot pairs located after the reference positions $\frac{\pi}{2}$ and $\frac{3\pi}{2}$, yielding the so-called "linear distribution", as illustrated in Fig. 1.18, with:

$$
\begin{cases}
N_{1-1'} = N_{6-6'} = \dfrac{N}{12} \\[2mm]
N_{2-2'} = N_{5-5'} = 2\dfrac{N}{12} \\[2mm]
N_{3-3'} = N_{4-4'} = 3\dfrac{N}{12}
\end{cases} \quad (1.71)
$$

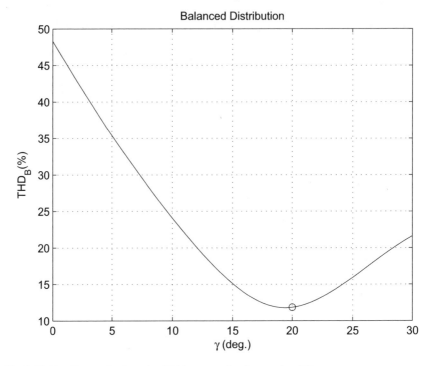

Fig. 1.17 Total harmonic distortion THD_B versus the slot angular shift γ

Fig. 1.18 Cross-section of a
synchronous machine
equipped by a single
coil-made field
linearly-distributed in six
slot pairs with $\beta = 1$

The spatial repartitions of the MMFs produced by the coils located in the six slot pairs and their resultant $\mathcal{F}_L$ are shown in Fig. 1.19.

Their *Fourier* expansions are derived taking into account their angular shifts with respect to the reference positions $\frac{\pi}{2}$ and $\frac{3\pi}{2}$, as follows:

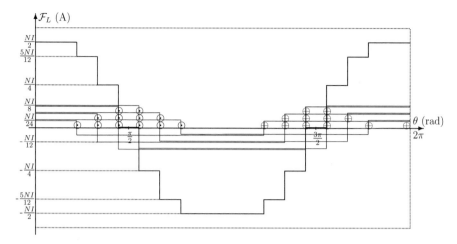

Fig. 1.19 Spatial repartitions of the MMFs created by the coils and their resultant $\mathcal{F}_L$

$$
\begin{cases}
\mathcal{F}_{1-1'}(\theta) = \sum_{n=0}^{n=\infty} \frac{2}{\pi} \frac{NI}{12} \frac{(-1)^n}{(2n+1)} \cos(2n+1)(\theta + 5\frac{\gamma}{2}) \\[2mm]
\mathcal{F}_{2-2'}(\theta) = \sum_{n=0}^{n=\infty} \frac{2}{\pi} \frac{NI}{6} \frac{(-1)^n}{(2n+1)} \cos(2n+1)(\theta + 3\frac{\gamma}{2}) \\[2mm]
\mathcal{F}_{3-3'}(\theta) = \sum_{n=0}^{n=\infty} \frac{2}{\pi} \frac{NI}{4} \frac{(-1)^n}{(2n+1)} \cos(2n+1)(\theta + \frac{\gamma}{2}) \\[2mm]
\mathcal{F}_{4-4'}(\theta) = \sum_{n=0}^{n=\infty} \frac{2}{\pi} \frac{NI}{4} \frac{(-1)^n}{(2n+1)} \cos(2n+1)(\theta - \frac{\gamma}{2}) \\[2mm]
\mathcal{F}_{5-5'}(\theta) = \sum_{n=0}^{n=\infty} \frac{2}{\pi} \frac{NI}{6} \frac{(-1)^n}{(2n+1)} \cos(2n+1)(\theta - 3\frac{\gamma}{2}) \\[2mm]
\mathcal{F}_{6-6'}(\theta) = \sum_{n=0}^{n=\infty} \frac{2}{\pi} \frac{NI}{12} \frac{(-1)^n}{(2n+1)} \cos(2n+1)(\theta - 5\frac{\gamma}{2})
\end{cases}
\tag{1.72}
$$

which could be partially summed as:

$$
\begin{cases}
\mathcal{F}_{1-1'}(\theta) + \mathcal{F}_{6-6'}(\theta) = \sum_{n=0}^{n=\infty} \frac{2}{\pi} \frac{NI}{6} \frac{(-1)^n}{(2n+1)} \cos(2n+1)5\frac{\gamma}{2} \cos(2n+1)\theta \\[2mm]
\mathcal{F}_{2-2'}(\theta) + \mathcal{F}_{5-5'}(\theta) = \sum_{n=0}^{n=\infty} \frac{2}{\pi} \frac{NI}{3} \frac{(-1)^n}{(2n+1)} \cos(2n+1)3\frac{\gamma}{2} \cos(2n+1)\theta \\[2mm]
\mathcal{F}_{3-3'}(\theta) + \mathcal{F}_{4-4'}(\theta) = \sum_{n=0}^{n=\infty} \frac{2}{\pi} \frac{NI}{2} \frac{(-1)^n}{(2n+1)} \cos(2n+1)\frac{\gamma}{2} \cos(2n+1)\theta
\end{cases}
\tag{1.73}
$$

Finally, the total sum leads to the *Fourier* expansion of the resultant MMF $\mathcal{F}_L(\theta)$ as:

$$\mathcal{F}_L(\theta) = \sum_{n=0}^{n=\infty} \frac{2}{\pi} \frac{NI}{6} \frac{(-1)^n}{(2n+1)} \left(\cos(2n+1)5\frac{\gamma}{2} + 2\cos(2n+1)3\frac{\gamma}{2} + 3\cos(2n+1)\frac{\gamma}{2} \right) \cos(2n+1)\theta$$

$$(1.74)$$

Discarding the harmonic of rank 3 is achieved by the slot angular shift γ fulfilling the following:

$$\cos 15\frac{\gamma}{2} + 2\cos 9\frac{\gamma}{2} + 3\cos 3\frac{\gamma}{2} = 0 \tag{1.75}$$

Its resolution gives $\gamma = \frac{\pi}{6} = 30°$.

In order to check the appropriateness of such a solution, let us compare it to the value corresponding to the minimum of the total harmonic distortion THD_L, as:

$$THD_L(\%) = 100 \frac{\sqrt{\sum_{n=1}^{n=\infty} \left(\frac{\cos(2n+1)5\frac{\gamma}{2} + 2\cos(2n+1)3\frac{\gamma}{2} + 3\cos(2n+1)\frac{\gamma}{2}}{2n+1} \right)^2}}{\cos 5\frac{\gamma}{2} + 2\cos 3\frac{\gamma}{2} + 3\cos \frac{\gamma}{2}} \tag{1.76}$$

The obtained results are shown in Fig. 1.20. One can notice that $\gamma = 30°$ does not correspond to the minimum of THD_L with $\min(THD_L) \simeq 13.41\%$. It is to be noted that this latter is higher than $\min(THD_B) \simeq 11.77\%$ which confirms the superiority of the balanced distribution with respect to the linear one in the case of six slot pairs.

1.3.2.3   Sinusoidal Distribution

The sinusoidal distribution is an improved version of the linear one. Indeed, the slot fill ratios are rethought in such a way that the number of turns per slot pair is proportional to the sinus of the angular shift measured with respect to the reference positions $\frac{\pi}{2}$ and $\frac{3\pi}{2}$, as follows:

$$\begin{cases} N_{1-1'} = N_{6-6'} = N_{sd} \sin\left(\frac{\pi - 5\gamma}{2} \right) \\ N_{2-2'} = N_{5-5'} = N_{sd} \sin\left(\frac{\pi - 3\gamma}{2} \right) \\ N_{3-3'} = N_{4-4'} = N_{sd} \sin\left(\frac{\pi - \gamma}{2} \right) \end{cases} \tag{1.77}$$

where N_{sd} is expressed as:

$$N_{sd} = \frac{N}{2\left(\sin\left(\frac{\pi - 5\gamma}{2} \right) + \sin\left(\frac{\pi - 3\gamma}{2} \right) + \sin\left(\frac{\pi - \gamma}{2} \right) \right)} \tag{1.78}$$

The *Fourier* expansions of the spatial repartitions of the MMFs created by the six coils located in the slot pairs: 1-1', 2-2', 3-3', 4-4', 5-5', and 6-6' are derived taking

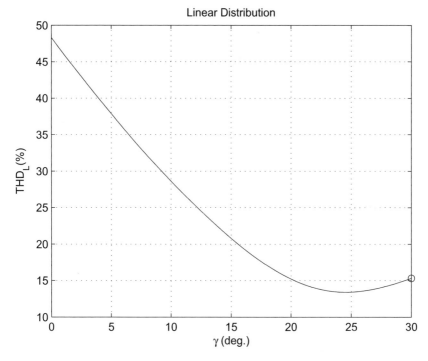

Fig. 1.20 Total harmonic distortion THD_L versus the slot angular shift γ

into account their angular shifts with respect to the reference positions $\frac{\pi}{2}$ and $\frac{3\pi}{2}$, as follows:

$$
\begin{cases}
\mathcal{F}_{1-1'}(\theta) = \sum_{n=0}^{n=\infty} \frac{2}{\pi} N_{sd} I \sin\left(\frac{\pi - 5\gamma}{2}\right) \frac{(-1)^n}{(2n+1)} \cos(2n+1)(\theta + 5\frac{\gamma}{2}) \\[2mm]
\mathcal{F}_{2-2'}(\theta) = \sum_{n=0}^{n=\infty} \frac{2}{\pi} N_{sd} I \sin\left(\frac{\pi - 3\gamma}{2}\right) \frac{(-1)^n}{(2n+1)} \cos(2n+1)(\theta + 3\frac{\gamma}{2}) \\[2mm]
\mathcal{F}_{3-3'}(\theta) = \sum_{n=0}^{n=\infty} \frac{2}{\pi} N_{sd} I \sin\left(\frac{\pi - \gamma}{2}\right) \frac{(-1)^n}{(2n+1)} \cos(2n+1)(\theta + \frac{\gamma}{2}) \\[2mm]
\mathcal{F}_{4-4'}(\theta) = \sum_{n=0}^{n=\infty} \frac{2}{\pi} N_{sd} I \sin\left(\frac{\pi - \gamma}{2}\right) \frac{(-1)^n}{(2n+1)} \cos(2n+1)(\theta - \frac{\gamma}{2}) \\[2mm]
\mathcal{F}_{5-5'}(\theta) = \sum_{n=0}^{n=\infty} \frac{2}{\pi} N_{sd} I \sin\left(\frac{\pi - 3\gamma}{2}\right) \frac{(-1)^n}{(2n+1)} \cos(2n+1)(\theta - 3\frac{\gamma}{2}) \\[2mm]
\mathcal{F}_{6-6'}(\theta) = \sum_{n=0}^{n=\infty} \frac{2}{\pi} N_{sd} I \sin\left(\frac{\pi - 5\gamma}{2}\right) \frac{(-1)^n}{(2n+1)} \cos(2n+1)(\theta - 5\frac{\gamma}{2})
\end{cases}
\tag{1.79}
$$

which could be partially summed as:

$$\begin{cases} \mathcal{F}_{1-1'}(\theta) + \mathcal{F}_{6-6'}(\theta) = \sum_{n=0}^{n=\infty} \frac{4}{\pi} N_{sd} I \sin\left(\frac{\pi - 5\gamma}{2}\right) \frac{(-1)^n}{(2n+1)} \cos(2n+1) 5\frac{\gamma}{2} \cos(2n+1)\theta \\[2mm] \mathcal{F}_{2-2'}(\theta) + \mathcal{F}_{5-5'}(\theta) = \sum_{n=0}^{n=\infty} \frac{4}{\pi} N_{sd} I \sin\left(\frac{\pi - 3\gamma}{2}\right) \frac{(-1)^n}{(2n+1)} \cos(2n+1) 3\frac{\gamma}{2} \cos(2n+1)\theta \quad (1.80) \\[2mm] \mathcal{F}_{3-3'}(\theta) + \mathcal{F}_{4-4'}(\theta) = \sum_{n=0}^{n=\infty} \frac{4}{\pi} N_{sd} I \sin\left(\frac{\pi - \gamma}{2}\right) \frac{(-1)^n}{(2n+1)} \cos(2n+1) \frac{\gamma}{2} \cos(2n+1)\theta \end{cases}$$

Finally, the total sum leads to the *Fourier* expansion of the resultant MMF $\mathcal{F}_S(\theta)$ as:

$$\mathcal{F}_S(\theta) = \sum_{n=0}^{n=\infty} F_{S(2n+1)} \cos(2n+1)\theta \tag{1.81}$$

where:

$$F_{S(2n+1)} = \frac{2}{\pi} NI \frac{(-1)^n}{(2n+1)} \left(\frac{\cos\frac{5\gamma}{2} \cos(2n+1)5\frac{\gamma}{2} + \cos\frac{3\gamma}{2}\cos(2n+1)3\frac{\gamma}{2} + \cos\frac{\gamma}{2}\cos(2n+1)\frac{\gamma}{2}}{\cos\frac{5\gamma}{2} + \cos\frac{3\gamma}{2} + \cos\frac{\gamma}{2}} \right) \tag{1.82}$$

The amplitude $F_{S(3)}$ of the harmonic of rank 3 is then expressed as follows:

$$F_{S3} = \frac{2}{\pi} NI \frac{(-1)^n}{(2n+1)} \left(\frac{\cos\frac{5\gamma}{2}\cos 15\frac{\gamma}{2} + \cos\frac{3\gamma}{2}\cos 9\frac{\gamma}{2} + \cos\frac{\gamma}{2}\cos 3\frac{\gamma}{2}}{\cos\frac{5\gamma}{2} + \cos\frac{3\gamma}{2} + \cos\frac{\gamma}{2}} \right) \tag{1.83}$$

It has been found that there are four values of γ that leads to $F_{S(3)} = 0$, as illustrated in Fig. 1.21 which shows the evolution of its normalized value $\widehat{F}_{S(3)}$, such that:

$$\widehat{F}_{S3} = \frac{\cos\frac{5\gamma}{2}\cos 15\frac{\gamma}{2} + \cos\frac{3\gamma}{2}\cos 9\frac{\gamma}{2} + \cos\frac{\gamma}{2}\cos 3\frac{\gamma}{2}}{\cos\frac{5\gamma}{2} + \cos\frac{3\gamma}{2} + \cos\frac{\gamma}{2}} \tag{1.84}$$

However, no value of γ among the four that enable the cancellation of $F_{S(3)}$ leads to an optimized distribution. Indeed, an investigation of the evolution of the total harmonic distortion THD_S with respect to γ has revealed that a minimum value is achieved by $\gamma \simeq 22°$, as shown in Fig. 1.22. For the sake of comparison, the curves giving THD_B and THD_L versus γ are recalled in Fig. 1.22. It has been found that the sinusoidal distribution exhibits the lowest value of the total harmonic distortion, that is $\min(THD_S) \simeq 11.73\%$.

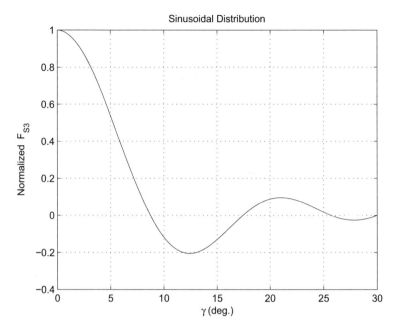

Fig. 1.21 Normalized amplitude of the harmonic of rank 3 versus the slot angular shift γ

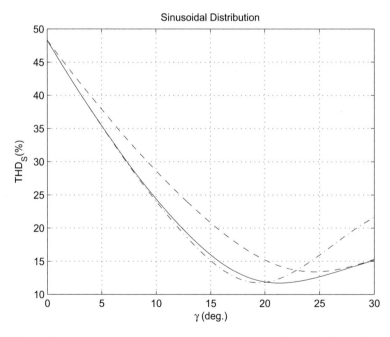

Fig. 1.22 Total harmonic distortion THD_S versus the slot angular shift γ in continuous line; THD_B and THD_L redrawn in mixed and dashed lines, respectively

1.4   Conclusion

This chapter dealt with the basis of the design of rotating AC machines with emphasis on their air gap magnetomotive force (MMF). To start with, The formulation of the MMF has been considered using the Amperes' theorem and the flux conservation law. Then, the harmonic content of its spatial repartition has been investigated. In a second part, different case studies have been treated including single and three phase concentrated windings. It has been found that the concentrated windings are penalized by the high harmonic content of their MMFs.

An attempt to reduce such harmonic content, based on the winding distribution, has been developed with the superposition approach-based formulation of the generated air gap MMFs. The potentialities of the balanced, linear, and sinusoidal distributions have been investigated and compared. It has been found that the sinusoidal distribution exhibits the lowest harmonic content.

In AC machines, the involved windings (single of multi-phase ones) are producing rotating fields on which their principle of operation lies. The following chapter is devoted to the creation of rotating fields in the air gap of AC machines and the resulting effects on their features.

Chapter 2
Rotating Fields: Creation and Effects on the Machine Features

Abstract The principle of operation of AC machines lies on the rotating field concept. Basically, it consists in a flux density moving in the cylindrical trajectory of the air gap of AC rotating machines. This chapter deals with the formulation of the rotating fields that could be generated considering different techniques, such as: (i) moving a single phase winding linked to the rotor and fed by a DC current (case of the field of the synchronous machines), and (ii) feeding a poly-phase winding by poly-phase sinusoidal currents. Then, the effects of the rotating field on the winding located in the vicinity, on the other side of the air gap, are analyzed. These effects are initiated by the induction of back-EMFs that could lead to the circulation of poly-phase currents in the involved winding. Consequently, a second rotating field is generated which is synchronized with the initial one, resulting in the production of an electromagnetic torque.

Keywords Rotating field · Single phase winding · Poly-phase winding
Synchronous speed · Back-EMFs · Electromagnetic torque

2.1 Introduction

The principle of rotating magnetic fields, or simply rotating fields, is the key to the operation of most AC rotating machines. Both synchronous and induction types machines rely on rotating fields in their air gaps. The discovery of the rotating field is generally attributed to two inventors, the Italian physicist and electrical engineer *Galileo Ferraris*, and the Serbian-American inventor and electrical engineer *Nikola Tesla*. In 1888 *Tesla* published a United States patent (U.S. Patent 0,381,968) for his invention related to rotating field while *Ferraris* published the results of his research on the same concept in a paper to the Royal Academy of Sciences in Turin.

Basically, a rotating field could be achieved by:

1. a single phase winding linked to the rotor and fed by a DC current. This technique is adopted in DC-excited synchronous machine. In this case, the involved single phase winding is called "the field". The same result could by achieved

© The Author(s) 2019
A. Masmoudi, *Design and Electromagnetic Feature Analysis*
of AC Rotating Machines, SpringerBriefs in Electrical and Computer
Engineering, https://doi.org/10.1007/978-981-13-0920-5_2

Fig. 2.1 A single phase
winding in the rotor achieved
by connecting in series p
concentrated coils of N turns
each characterized by
$\alpha = \tau_p = \frac{\pi}{p} \; (\beta = 1)$

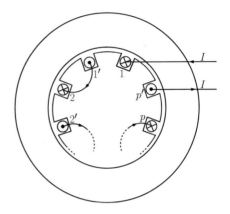

by substituting the field by a set of permanent magnet pair(s). If the rotor is
moving at a speed Ω, the field (or the permanent magnets) would rotate at the
same speed, resulting in the creation of a rotating field,

2. a poly-phase winding, linked to the stator of the rotor, which is fed by poly-
phase sinusoidal currents with their phase shifts equal to the electrical angular
displacements between the phases. In this case, the speed of the rotating field is
equal to the angular frequency of the currents.

Rotating fields are affecting the windings located on the other side of the air gap by
inducing in back-EMFs. These could be accessed under no-load generator operation
of the machine. Now, if the winding with induced back-EMFs is closed through a
power supply or a load, it would generate a second rotating field. The synchronisation
between the initial rotating field and the one generated by the winding represents an
approach to develop an electromagnetic torque.

The chapter is aimed at the different techniques enabling the production of rotating
fields considering the case of a single phase winding linked to the rotor and fed by
a DC current and the case of a poly-phase winding fed by poly-phase sinusoidal
currents. Then, the effects of the created rotating fields on the windings located on
the other side of the air gap, are analyzed.

2.2 Rotating Field Production

2.2.1 Case of a Single Phase Winding in the Rotor

Let us consider the case of a single phase winding made up of p concentrated coils
of N turns each. These are inserted in $2p$ slots with a slot pitch α equal to the pole
pitch $\tau_p = \frac{\pi}{p}$, as illustrated in Fig. 2.1.

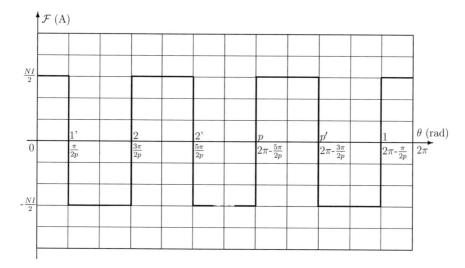

Fig. 2.2 Spatial repartition of the MMF created by a single phase winding in the rotor

Feeding the winding by a DC current I results in the creation of a MMF in the air gap exhibiting the spatial repartition shown in Fig. 2.2 where the magnetic axis of the coil inserted in the slot pair 1-1' is taken as the origin of the mechanical angle θ.

Referring to Fig. 2.2, one can notice that the spatial repartition of the MMF presents p periods which leads to the following relation between the electrical angle θ_e and the mechanical one θ:

$$\theta_e = p\theta \tag{2.1}$$

Its *Fourier* expansion is expressed as follows:

$$\mathcal{F}(\theta) = \frac{2}{\pi}NI\left(\cos p\theta - \frac{1}{3}\cos 3p\theta + \frac{1}{5}\cos 5p\theta + \ldots\ldots + \frac{(-1)^n}{(2n+1)}\cos(2n+1)p\theta\right) \tag{2.2}$$

Such MMF varies with the space (a long the air gap) but not with the time.

Now, let us suppose that the rotor is driven at a constant speed equal to n turns per second. This leads to a displacement of the MMF waveform at the speed v (m/s), with:

$$v = \pi Dn \tag{2.3}$$

where D is the air gap diameter.

Hence, an observer linked to the stator will record the scrolling of pn periods of the MMF per second. That is to say, a stator winding will be affected by the variable flux. Consequently, it will generate a back-EMF with a fundamental characterized by an angular frequency ω, such that:

$$\omega = 2\pi \, pn \qquad (2.4)$$

Accounting for Eq. (2.4), the expression of the speed v can be rewritten as:

$$v = \frac{\pi D}{p} \frac{\omega}{2\pi} \qquad (2.5)$$

which is the propagation speed of an electromagnetic wave that has an angular frequency ω and a wave length λ, with:

$$\lambda = \frac{\pi D}{p} \qquad (2.6)$$

Thus, for a given point M in the air gap marked by its position x with respect to a magnetic axis taken as the angle frame, and for an instant t, one can express the fundamental flux density as follows:

$$b_1(x, t) = B_1 \cos\left(2\pi\frac{x}{\lambda} - \omega t\right) \qquad (2.7)$$

where:

$$B_1 = \frac{\mu_0}{e}\frac{2}{\pi}NI \qquad (2.8)$$

Given that:

$$x = \frac{D}{2}\theta \qquad (2.9)$$

and that the rotor speed Ω (rad/s) is expressed as:

$$\Omega = 2\pi n = \frac{\omega}{p} \qquad (2.10)$$

Equation (2.11) can be rewritten in terms θ and t as follows:

$$b_1(\theta, t) = B_1 \cos\left(p(\theta - \Omega t)\right) \qquad (2.11)$$

This represents the expression of the so-called "rotating field" which is a flux density wave rotating in the air gap at the same speed as the rotor Ω.

In the manner of the fundamental, the harmonics create rotating fields turning at the same speed Ω, such as:

$$\begin{cases} b_3(x,t) & = B_3 \cos\left(3p(\theta - \Omega t)\right) \\ b_5(x,t) & = B_5 \cos\left(5p(\theta - \Omega t)\right) \\ \text{.......} & \text{.. ....} \\ b_{(2n+1)}(x,t) & = B_{(2n+1)} \cos\left((2n+1)p(\theta - \Omega t)\right) \end{cases} \qquad (2.12)$$

However, they will induce in the stator winding back-EMFs that have as angular frequencies 3Ω, 5ω, ..., $(2n+1)\omega$, respectively. In other words, the resultant back-EMF has an harmonic content which is directly linked to the one of the rotating field. Therefore, in order to produce nearly sinusoidal back-EMFs, a great attention has to be paid to the design of the winding creating the air gap MMF, as already treated in the previous chapter.

Remarks Some remarks may arise at this level, as:

- If the winding creating the rotating field is kept in the rotor, and if the stator winding is transferred to the rotor, no back-EMF is induced in this later because the rotating field turns to be immobile with respect to it,
- if the rotor winding is transferred to the stator and is fed by the DC current I. The air gap flux density is no longer a rotating filed. Consequently:

  - there is no back-EMF induced in the second winding if it would be kept in the stator,
  - there is a back-EMF induced in the second winding if it would be transferred to the rotor.

2.2.2 Case of a Three Phase Winding

Let us consider the case of a three phase winding which could be located in the rotor or in the stator. For the sake of simplicity, the case of a three phase winding, made up of a single coil of N turns concentrated in a slot pair per phase with $\alpha = \tau_p = \pi$, is considered. Such a winding is shown in Fig. 2.3.

Let us:

- call (i_a, i_b, i_c) the currents circulating in phases "a", "b", and "c", respectively,
- consider a point M in the air gap located in a position θ marked with respect to the magnetic axis of phase "a".

The fundamental flux density $b_1(\theta)$ in point M is the superposition of the ones produced by phases "a", "b", and "c", such that:

$$\begin{cases} b_{1a}(\theta) = \frac{\mu_0}{e} \frac{2}{\pi} N i_a \cos\theta \\ b_{1b}(\theta) = \frac{\mu_0}{e} \frac{2}{\pi} N i_b \cos\left(\theta - \frac{2\pi}{3}\right) \\ b_{1c}(\theta) = \frac{\mu_0}{e} \frac{2}{\pi} N i_c \cos\left(\theta + \frac{2\pi}{3}\right) \end{cases} \qquad (2.13)$$

which leads to:

Fig. 2.3 Cross-section of a machine equipped with a three phase winding in the stator

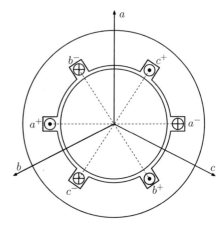

$$b_1(\theta) = \frac{\mu_0}{e} \frac{2}{\pi} N \left(i_a \cos\theta \ + \ i_b \cos\left(\theta - \frac{2\pi}{3}\right) \ + \ i_c \cos\left(\theta + \frac{2\pi}{3}\right) \right) \quad (2.14)$$

Let us assume that $i_a = i_b = i_c = i$. In this case, Eq. (2.14) turns to be:

$$b_1(\theta) = \frac{\mu_0}{e} \frac{2}{\pi} Ni \left[\cos\theta \ + \ \cos\left(\theta - \frac{2\pi}{3}\right) \ + \ \cos\left(\theta + \frac{2\pi}{3}\right) \right] = 0 \ \ \forall \theta \quad (2.15)$$

which is not the case of the harmonic of rank 3 of the flux density $b_3(\theta)$, whose formulation considering the same approach as for the fundamental, yields:

$$b_3(\theta) = -\frac{1}{3} \frac{\mu_0}{e} \frac{2}{\pi} Ni \left[\cos 3\theta + \cos\left(3(\theta - \frac{2\pi}{3})\right) + \cos\left(3(\theta + \frac{2\pi}{3})\right) \right] = -\frac{\mu_0}{e} \frac{2}{\pi} Ni \cos 3\theta$$
$$(2.16)$$

Now let us consider the case where:

$$\begin{cases} i_a = \sqrt{2} I_{rms} \cos\omega t \\ i_b = \sqrt{2} I_{rms} \cos\left(\omega t - \frac{2\pi}{3}\right) \\ i_c = \sqrt{2} I_{rms} \cos\left(\omega t + \frac{2\pi}{3}\right) \end{cases} \quad (2.17)$$

The substitution of i_a, i_b, and i_c by their expressions in Eq. (2.14) yields:

$$b_1(\theta) = \frac{\mu_0}{e} \frac{2}{\pi} N\sqrt{2} I_{rms} \left(\cos\theta \cos\omega t + \cos(\theta - \frac{2\pi}{3})\cos(\omega t - \frac{2\pi}{3}) + \cos(\theta + \frac{2\pi}{3})\cos(\omega t + \frac{2\pi}{3}) \right) \quad (2.18)$$

that gives:

$$b_1(\theta) = \frac{\mu_0}{e} \frac{2}{\pi} N\sqrt{2} I_{rms} \frac{3}{2} \cos(\theta - \omega t) \quad (2.19)$$

Comparing Eqs. (2.11) and (2.19), one can notice that this later represents a rotating field moving in the air gap at a speed equal to the angular frequency ω of the currents feeding the three phases and for a pole pair $p = 1$.

Now let us extend the study to the space harmonics (harmonics of the spatial repartition of the flux density), a similar derivation has led to:

$$\begin{cases} b_5(\theta) & = \frac{\mu_0}{e}\frac{2}{\pi}N\sqrt{2}I_{rms}\frac{3}{2}\left(\frac{1}{5}\right) & \cos(5\theta - \omega t) \\ b_7(\theta) & = \frac{\mu_0}{e}\frac{2}{\pi}N\sqrt{2}I_{rms}\frac{3}{2}\left(-\frac{1}{7}\right) & \cos(7\theta - \omega t) \\ ..... & \quad .. \quad .... & \quad .... \\ b_{(2n+1)}(\theta) & = \frac{\mu_0}{e}\frac{2}{\pi}N\sqrt{2}I_{rms}\frac{3}{2}\left(\frac{(-1)^n}{(2n+1)}\right) & \cos((2n+1)\theta - \omega t) \end{cases} \qquad (2.20)$$

In the manner of the fundamental, the harmonics produce rotating fields at the speed ω.

Pulsating Field Let us assume that just the "a" phase is fed by the current i_a. By limiting the analysis to the fundamental, the air gap flux density is expressed as:

$$b_1(\theta) = \frac{\mu_0}{e}\frac{2}{\pi}N\sqrt{2}I_{rms}\cos\theta\cos\omega t \qquad (2.21)$$

which represents the expression of a pulsating field moving up a down within the magnetic axis of the "a" phase with $b_1(\frac{\pi}{2}) = b_1(\frac{3\pi}{2}) = 0\text{T}$ for all instants. Actually, Eq. (2.21) could be rewritten as:

$$b_1(\theta) = \frac{\mu_0}{e}\frac{2}{\pi}N\sqrt{2}I_{rms}\frac{1}{2}(\cos(\theta - \omega t) + \cos(\theta + \omega t)) \qquad (2.22)$$

which is the sum of two rotating fields moving in the air gap at the same speed ω but in opposite direction.

2.2.3 Case of a Two Phase Winding

Basically, it consists in a winding made up of four phases "a", "b", "c" and "d" of N turns each, which are shifted by 90° electrical angle and are fed by:

$$\begin{cases} i_a = \sqrt{2}I_{rms}\cos\omega t \\ i_b = \sqrt{2}I_{rms}\cos\left(\omega t - \frac{\pi}{2}\right) \\ i_c = \sqrt{2}I_{rms}\cos(\omega t - \pi) \\ i_d = \sqrt{2}I_{rms}\cos\left(\omega t - \frac{3\pi}{2}\right) \end{cases} \qquad (2.23)$$

Figure 2.4 shows a machine equipped by a four phase winding in the stator ($p = 1$).

Fig. 2.4 Cross-section of a machine equipped by a four phase winding in the stator ($p = 1$)

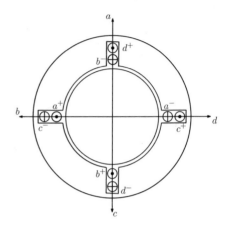

Accounting for the following remarks:

- the coils of phases "a" and "c" are inserted in the same slots and the coils of phases "b" and "d" are inserted in the same slots,
- the magnetic axis of phases "a" and "c" are opposite and the magnetic axis of phases "b" and "d" are opposite,
- the currents i_a and i_c are opposite and the currents i_b and i_d are opposite,

phases "a" and "c" are merged and turn to be a single phase named "a" and phases "b" and "d" are merged and turn to be a single phase named "b". Thus, the winding is reduced to two phases which are shifted by 90° electrical angle and are fed by:

$$\begin{cases} i_a = \sqrt{2}I_{rms}\cos\omega t \\ i_b = \sqrt{2}I_{rms}\cos\left(\omega t - \frac{\pi}{2}\right) \end{cases} \tag{2.24}$$

Let us consider a point M in the air gap located in a position θ marked with respect to the magnetic axis of phase "a", and by limiting the study to the fundamental terms, the fundamental flux density $b_1(\theta)$ in point M is the superposition of the ones produced by phases "a" and "b", as:

$$b_1(\theta) = b_{1a}(\theta) + b_{1b}(\theta) = \frac{\mu_0}{e}\frac{2}{\pi}N\left(i_a\cos\theta + i_b\cos\left(\theta - \frac{\pi}{2}\right)\right) \tag{2.25}$$

The substitution of i_a and i_b by their expressions in Eq. (2.25) leads to:

$$b_1(\theta) = \frac{\mu_0}{e}\frac{2}{\pi}N\sqrt{2}I_{rms}\left(\cos\theta\cos\omega t + \cos\left(\theta - \frac{\pi}{2}\right)\cos\left(\omega t - \frac{\pi}{2}\right)\right) \tag{2.26}$$

which is simply equal to:

$$b_1(\theta) = \frac{\mu_0}{e}\frac{2}{\pi}N\sqrt{2}I_{rms}\left(\cos\theta\cos\omega t + \sin\theta\sin\omega t\right) \tag{2.27}$$

and finally:

$$b_1(\theta) = \frac{\mu_0}{e} \frac{2}{\pi} N \sqrt{2} I_{rms} \cos(\theta - \omega t) \tag{2.28}$$

which is the expression of a rotating field moving in the air gap at a speed equal to the current angular frequency ω.

In the manner of the fundamental, the space harmonics produce rotating fields at the same speed ω which are expressed as follows:

$$\begin{cases} b_3(\theta) & = \frac{\mu_0}{e} \frac{2}{\pi} N \sqrt{2} I_{rms} \left(\frac{-1}{3}\right) & \cos(3\theta + \omega t) \\ b_5(\theta) & = \frac{\mu_0}{e} \frac{2}{\pi} N \sqrt{2} I_{rms} \left(\frac{1}{5}\right) & \cos(5\theta - \omega t) \\ \cdots & \cdots & \cdots \\ b_{(2n+1)}(\theta) & = \frac{\mu_0}{e} \frac{2}{\pi} N \sqrt{2} I_{rms} \left(\frac{(-1)^n}{(2n+1)}\right) & \cos((2n+1)\theta - \omega t) \end{cases} \tag{2.29}$$

Remark The principle of operation of the single phase induction motor lies on the above-described concept. The single phase induction motor is equipped by a squirrel cage in the rotor and by two phases in the stator which are shifted by 90° electrical angle, with:

- one phase directly linked to a single phase grid,
- the other phase fed by the same grid through a capacitor in an attempt to achieve a $\frac{\pi}{2}$ shift of its current with respect to the one of the previous phase.

The denomination "single phase" is related to the motor supply and not to its topology which is actually a two phase one.

2.2.4 Case of a Five Phase Winding

It is a winding made up of five phases "a", "b", "c", "d" et "e" which are shifted by an electrical angle equal to 72°. Let us consider the case where the phases are concentrated in five slot pairs of N turns each and are fed by sinusoidal currents, as:

$$\begin{cases} i_a = \sqrt{2} I_{rms} \cos \omega t \\ i_b = \sqrt{2} I_{rms} \cos \left(\omega t - \frac{2\pi}{5}\right) \\ i_c = \sqrt{2} I_{rms} \cos \left(\omega t - \frac{4\pi}{5}\right) \\ i_d = \sqrt{2} I_{rms} \cos \left(\omega t - \frac{6\pi}{5}\right) \\ i_e = \sqrt{2} I_{rms} \cos \left(\omega t - \frac{8\pi}{5}\right) \end{cases} \tag{2.30}$$

The fundamental flux density $b_1(\theta)$ in a point M in the air gap marked by its position θ $(p = 1)$ with respect to the magnetic axis of phase "a" is expressed as follows:

$$b_1(\theta) = \frac{\mu_0}{e}\frac{2}{\pi}N\left(i_a\cos\theta + i_b\cos\left(\theta - \frac{2\pi}{5}\right)\right.$$

$$\left. +i_c\cos\left(\theta - \frac{4\pi}{5}\right) + i_d\cos\left(\theta - \frac{6\pi}{5}\right) + i_e\cos\left(\theta - \frac{8\pi}{5}\right)\right) \quad (2.31)$$

The substitution of the currents by their expressions in Eq. (2.31) leads to:

$$b_1(\theta,t) = \frac{\mu_0}{e}\frac{2}{\pi}N\sqrt{2}I_{rms}\left[\cos\theta\cos\omega t + \cos\left(\theta - \frac{2\pi}{5}\right)\cos\left(\omega t - \frac{2\pi}{5}\right)\right.$$

$$\left. +\cos\left(\theta - \frac{4\pi}{5}\right)\cos\left(\omega t - \frac{4\pi}{5}\right) + \cos\left(\omega t - \frac{6\pi}{5}\right)\cos\left(\theta - \frac{6\pi}{5}\right) + \cos\left(\omega t - \frac{8\pi}{5}\right)\cos\left(\theta - \frac{8\pi}{5}\right)\right]$$

which is developed as:

$$b_1(\theta,t) = \frac{\mu_0}{e}\frac{2}{\pi}N\sqrt{2}I_{rms}\left[\left(\cos\theta\cos\frac{2\pi}{5} + \sin\theta\sin\frac{2\pi}{5}\right)\left(\cos\omega t\cos\frac{2\pi}{5} + \sin\omega t\sin\frac{2\pi}{5}\right)\right.$$

$$+\left(\cos\theta\cos\frac{4\pi}{5} + \sin\theta\sin\frac{4\pi}{5}\right)\left(\cos\omega t\cos\frac{4\pi}{5} + \sin\omega t\sin\frac{4\pi}{5}\right)$$

$$+\left(\cos\theta\cos\frac{6\pi}{5} + \sin\theta\sin\frac{6\pi}{5}\right)\left(\cos\omega t\cos\frac{6\pi}{5} + \sin\omega t\sin\frac{6\pi}{5}\right)$$

$$\left. +\left(\cos\theta\cos\frac{8\pi}{5} + \sin\theta\sin\frac{8\pi}{5}\right)\left(\cos\omega t\cos\frac{8\pi}{5} + \sin\omega t\sin\frac{8\pi}{5}\right) + \cos\theta\cos\omega t\right] \quad (2.32)$$

Giving that:

$$\begin{cases}\dfrac{6\pi}{5} = 2\pi - \dfrac{4\pi}{5} \\[2mm] \dfrac{8\pi}{5} = 2\pi - \dfrac{2\pi}{5}\end{cases} \quad (2.33)$$

the expression of $b_1(\theta, t)$ turns to be:

$$b_1(\theta,t) = \frac{\mu_0}{e}\frac{2}{\pi}N\sqrt{2}I_{rms}\left[\left(1 + 2\left(\cos\frac{2\pi}{5}\right)^2 + 2\left(\cos\frac{4\pi}{5}\right)^2\right)\cos\theta\cos\omega t\right.$$

$$\left. +\left(2\left(\sin\frac{2\pi}{5}\right)^2 + 2\left(\sin\frac{4\pi}{5}\right)^2\right)\sin\theta\sin\omega\right] \quad (2.34)$$

which is equal to:

$$b_1(\theta,t) = \frac{\mu_0}{e}\frac{2}{\pi}N\sqrt{2}I_{rms}\left[(2 + A_5)\cos\theta\cos\omega t + (3 - A_5)\sin\theta\sin\omega\right] \quad (2.35)$$

where:

$$A_5 = 2\left(\cos\frac{2\pi}{5}\right)^2 + \cos\frac{2\pi}{5} = 0.5 \tag{2.36}$$

that leads to the expression of a rotating field at a speed equal to the current angular frequency ω, as:

$$b_1(\theta, t) = \frac{\mu_0}{e}\frac{2}{\pi}N\sqrt{2}I_{rms}\frac{5}{2}\cos(\theta - \omega t) \tag{2.37}$$

In the manner of the fundamental, the space harmonics produce rotating fields at the same speed ω which are expressed as follows:

$$\begin{cases} b_3(\theta) & = \frac{\mu_0}{e}\frac{2}{\pi}N\sqrt{2}I_{rms}\frac{5}{2}\left(\frac{-1}{3}\right) & \cos(3\theta - \omega t) \\ b_7(\theta) & = \frac{\mu_0}{e}\frac{2}{\pi}N\sqrt{2}I_{rms}\frac{5}{2}\left(\frac{-1}{7}\right) & \cos(7\theta - \omega t) \\ \cdots\cdots & \cdots\cdots\cdots\cdots\cdots & \cdots\cdots\cdots \\ b_{(2n+1)}(\theta) & = \frac{\mu_0}{e}\frac{2}{\pi}N\sqrt{2}I_{rms}\frac{5}{2}\left(\frac{(-1)^n}{(2n+1)}\right) & \cos((2n+1)\theta - \omega t) \end{cases} \tag{2.38}$$

2.2.5 Generalisation

Let us consider a winding made up of q phases (with q an odd integer higher than unity) which are shifted by a electrical angle equal to $\frac{2\pi}{q}$. Let us consider the case where the phases are concentrated in q slot pairs of N turns each and are fed by sinusoidal currents. These have a rms-value I_{rms} and an angular frequency ω and are shifted by $\frac{2\pi}{q}$. The resulting fundamental rotating field is expressed as follows:

$$b_1(\theta, t) = \frac{\mu_0}{e}\frac{2}{\pi}N\sqrt{2}I_{rms}\frac{q}{2}\cos(\theta - \omega t) \tag{2.39}$$

Il appears at a first glance that increasing the number of phases leads to an increase of the magnitude of the rotating field. This is true if the magnetic circuit is not saturated which is practically unusual. Actually, the benefit of increasing the number of phases is the subdivision of the power in q circuits so that in case of a failure of one among them, the missed power is equal $\frac{1}{q}$ of the total power.

2.3 Back-EMF Induction

Rotating fields, characterizing the operation of AC machines, affect the windings located on the other side of the air gap "moving" at different speeds by inducting in back-EMFs. This is achieved in:

Fig. 2.5 Principle of the
back-EMF induction in a
concentrated coil

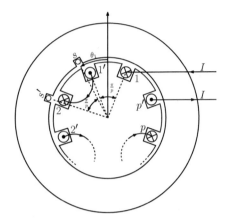

- synchronous machines where the rotating field is created by the field or the permanent magnets in the rotor and the back-EMFs are induced in the armature in the stator,
- induction machine where the rotating field at a speed Ω is created by the three phase winding in the stator and the back-EMFs are induced in the rotor circuits moving at a speed $\Omega_m < \Omega$.

2.3.1 Case of a Concentrated Coil

Let us consider the case of the synchronous machine where:

- the field is achieved by a single phase winding made up of p concentrated coils of N turns each. These are inserted in $2p$ slots with a slot pitch α equal to the pole pitch $\tau_p = \frac{\pi}{p}$,
- a concentrated coil of N_c is inserted in a slot pair s-s' in the stator, with a slot pitch equal to the rotor pole pitch τ_p, as illustrated in Fig. 2.5.

Let us consider the fundamental rotating field created by the field, expressed as:

$$b_1(\theta, t) = B_1 \cos\left(p(\theta - \Omega t)\right) = B_1 \cos\left(p\theta - \omega t\right) \tag{2.40}$$

where θ is a mechanical angle counted with respect of the position of the magnetic axis of one coil of the field at the initial time.

Let us call θ_1 the position of the stator slot s with respect to the same magnetic axis. During dt, the rotating field has moved with an angle $d\theta = \Omega dt$, and the flux linkage $d\Phi$ of a given conductor located in slot s is expressed as:

$$d\Phi_s = dS \, B_1 \cos\left(p\theta_1 - \omega t\right) \tag{2.41}$$

where:

$$dS = L_s \frac{D}{2} d\theta \tag{2.42}$$

with L_s and D are the stack length and the air gap diameter, respectively.

The back-EMF induced in the conductor is given by the *Faraday* law, such that:

$$e_s = -\frac{d\Phi_s}{dt} = -L\frac{D}{2}\Omega B_1 \cos(p\theta_1 - \omega t) \tag{2.43}$$

Now, let us consider one turn inserted in the slot pair s-s'. Only the two conductors located in the slots will have induced back-EMFs, such that:

$$\begin{cases} e_s = -L\frac{D}{2}\Omega B_1 \cos(p\theta_1 - \omega t) \\ e_{s'} = -L\frac{D}{2}\Omega B_1 \cos\left(p(\theta_1 + \frac{\pi}{p}) - \omega t\right) \end{cases} \tag{2.44}$$

that gives a back-EMF e_t induced in one turn, as follows:

$$e_t = e_{s'} - e_s = LD\Omega B_1 \cos(p\theta_1 - \omega t) \tag{2.45}$$

Finally, the back-EMF e_c induced in the stator concentrated winding is expressed as:

$$e_c = N_s e_t = N_c LD\Omega B_1 \cos(p\theta_1 - \omega t) \tag{2.46}$$

One can notice that e_c is sinusoidal because the rotating field has been assumed to be so (just the fundamental is taken into consideration in the above formulation), with:

- an angular frequency ω linked to the speed of the rotating field, such as $\omega = p\Omega$,
- an amplitude depending on the machine geometry (L and D), the number of turns, and the speed of the rotating field. Whereas the flux density amplitude B_1 is more or less affecting the back-EMF one in so far as it is limited by the saturation of the magnetic circuit.

2.3.2 Case of a Distributed Coil

2.3.2.1 Case Study

Now, let us suppose that the N_c turns of the coil are distributed in m slot pairs around the positions θ_1 and $\theta_1 + \frac{\pi}{p}$, which are shifted by an angle γ. For instance, Fig. 2.6 shows the case where $m = 3$.

Fig. 2.6 Principle of the
back-EMF induction in a
distributed coil (balanced
distribution)

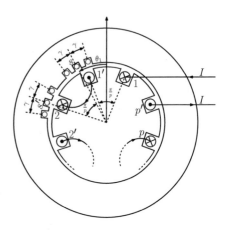

If the N_s turns are distributed in a balanced way in the three slot pairs, then one
can write:

$$\begin{cases} e_{s(1-1')} = \frac{N_c}{3} LD\Omega B_1 \cos\left(p(\theta_1 - \gamma) - \omega t\right) \\ e_{s(2-2')} = \frac{N_c}{3} LD\Omega B_1 \cos\left(p\theta_1 - \omega t\right) \\ e_{s(3-3')} = \frac{N_c}{3} LD\Omega B_1 \cos\left(p(\theta_1 + \gamma) - \omega t\right) \end{cases} \tag{2.47}$$

The total back-EMF induced in the coil distributed in three slot pairs is expressed
as:

$$e_d = \frac{(1 + 2\cos p\gamma)}{3} N_s LD\Omega B_1 \cos\left(p\theta_1 - \omega t\right) \tag{2.48}$$

It is to be noted that e_d has the same angular frequency and initial phase as e_c,
however its amplitude E_d is lower than E_c, with:

$$K_{d1} = \frac{E_d}{E_c} = \frac{1 + 2\cos p\gamma}{3} \tag{2.49}$$

K_{d1} is the so-called "distribution factor" of the fundamental back-EMF.

2.3.2.2 Generalization

Back-EMF Fundamental Let us suppose that the stator is equipped by q phases
of $2p$ poles and N_p turns each. These phases are distributed regularly (balanced
distribution) in N_s slots. Let us call m the number of slot(s) per pole and per phase,
such that:

Fig. 2.7 Back-EMF phasor diagram

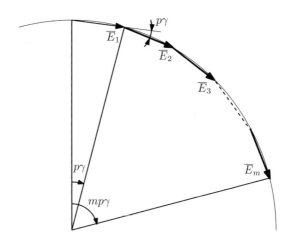

$$m = \frac{N_s}{2pq} \tag{2.50}$$

The fundamental back-EMF induced in one coil among the p ones of a phase is the sum of m back-EMFs which have an amplitude equal to $\frac{E_c}{m}$ and are shifted by $p\gamma = p\frac{2\pi}{N_s}$. These could be represented by a phasor diagram, as shown in Fig. 2.7 where $\overline{E}_1, \overline{E}_2 \ldots \overline{E}_m$ are the vectors representing the back-EMFs induced in the m equal parts of the distributed coil.

Referring to Fig. 2.7, the extremities of $\overline{E}_1, \overline{E}_2 \ldots \overline{E}_m$ are located on a circle with a radius $\mathcal{R}$, such that:

$$\mathcal{R} = \frac{\frac{E}{2}}{\sin\frac{p\gamma}{2}} \tag{2.51}$$

where:

$$E = \| \overline{E}_1 \| = \| \overline{E}_2 \| = \ldots\ldots\ldots\ldots = \| \overline{E}_m \| \tag{2.52}$$

The vectorial sum $\overline{E}_1 + \overline{E}_2 + \ldots\ldots\ldots + \overline{E}_m$ leads to the vector $\overline{E}_d$ representing the fundamental back-EMF induced in the coil.

Referring to Fig. 2.7, one can establish the following expression of $\mathcal{R}$:

$$\mathcal{R} = \frac{\frac{\| \overline{E}_d \|}{2}}{\sin\frac{pm\gamma}{2}} \tag{2.53}$$

The equality of both expressions of $\mathcal{R}$ given by Eqs. (2.51) and (2.53) gives:

$$\frac{\frac{E}{2}}{\sin \frac{p\gamma}{2}} = \frac{\frac{\| \overline{E_d} \|}{2}}{\sin \frac{pm\gamma}{2}} \qquad (2.54)$$

Giving that:

$$K_{d1} = \frac{\| \overline{E_d} \|}{m E} \qquad (2.55)$$

one can establish the expression of the distribution factor of the fundamental back-EMF, as follows:

$$K_{d1} = \frac{\sin \frac{pm\gamma}{2}}{m \sin \frac{p\gamma}{2}} \qquad (2.56)$$

Giving that:

$$pm\gamma = p \frac{2\pi}{N_s} \frac{N_s}{2pq} = \frac{\pi}{q} \qquad (2.57)$$

the expression of the distribution factor turns to be:

$$K_{d1} = \frac{\sin \frac{\pi}{2q}}{m \sin \frac{\pi}{2qm}} \qquad (2.58)$$

The fundamental back-EMF per phase is expressed as:

$$e_p = p K_{d1} N_c L D \Omega B_1 \cos(p\theta_1 - \omega t) \qquad (2.59)$$

Harmonic Back-EMFs The space harmonics of the rotating field induce in the phase harmonic back-EMFs of ranks $(2n + 1)$ with $n > 0$, such as:

$$e_{p(2n+1)} = p K_{d(2n+1)} N_c L D \Omega B_{(2n+1)} \cos((2n + 1)(p\theta_1 - \omega t)) \qquad (2.60)$$

where:

$$K_{d(2n+1)} = \frac{\sin \frac{(2n + 1)pm\gamma}{2}}{m \sin \frac{(2n + 1)p\gamma}{2}} = \frac{\sin \frac{(2n + 1)\pi}{2q}}{m \sin \frac{(2n + 1)\pi}{2qm}} \qquad (2.61)$$

One can notice that the angular frequency $\omega_{(2n+1)}$ of the induced harmonic back-EMFs of rank $(2n + 1)$ is a function of the fundamental one, as:

$$\omega_{(2n+1)} = (2n + 1)\omega \qquad (2.62)$$

Table 2.1 Distribution factors of the fundamental and harmonics of the back-EMF induced in a three phase winding regularly distributed in slot pairs characterized by a slot pitch equal to the pole pitch for different values of the slot(s) per pole and per phase m

	$m = 1$	$m = 2$	$m = 3$	$m = 4$	$m = 5$	$m \to \infty$
k_{d1}	1	0.9659	0.9598	0.9577	0.9567	$\to \frac{3}{\pi} \simeq 0.9549$
k_{d3}	1	0.7071	0.6667	0.6533	0.6472	$\to \frac{2}{\pi} \simeq 0.6366$
k_{d5}	1	0.2588	0.2176	0.2053	0.2000	$\to \frac{3}{5\pi} \simeq 0.1910$
$\lvert k_{d7} \rvert$	1	0.2588	0.1774	0.1576	0.1494	$\to \frac{3}{7\pi} \simeq 0.1364$
$\lvert k_{d9} \rvert$	1	0.7071	0.3333	0.2706	0.2472	$\to \frac{2}{3\pi} \simeq 0.2122$
$\lvert k_{d11} \rvert$	1	0.9659	0.1774	0.1261	0.1095	$\to \frac{3}{11\pi} \simeq 0.0868$

Case of Three Phase Machines The distribution factors of the fundamental and the harmonics of the induced back-EMF in a phase are obtained by substituting q by 3 in Eqs. (2.58) and (2.61). The resulting expressions are as follows:

$$
\begin{cases}
K_{d1} = \dfrac{1}{2m \sin \dfrac{\pi}{6m}} \\[3ex]
K_{d(2n+1)} = \dfrac{\sin \dfrac{(2n+1)\pi}{6}}{m \sin \dfrac{(2n+1)\pi}{6m}}
\end{cases}
\tag{2.63}
$$

The application of the expressions of the distribution factors given in Eq. (2.63) for different value has led to the results classified in Table 2.1.

Referring to Table 2.1, one can notice that distributing the winding leads to:

- a decrease of the amplitude of its fundamental back-EMF. Indeed, a decrease of the distribution factor K_{d1} from 1 to 0.9549 is noticed with the increase of m,
- a decrease followed by an increase of the distribution factor of its harmonic back-EMFs reaching the one of the fundamental. For instance, if $m = 2$, except for the harmonics multiple of three which are discarded in three phase machines, one can find the ones exhibiting the same distributing winding as the fundamental by writing the equality of K_{d1} and $\lvert K_{d(2n+1)} \rvert$ in Eq. (2.63). This gives:

$$
\frac{(2n+1)\pi}{12} = \frac{\pi}{12} + 2k\pi
\tag{2.64}
$$

leading to:

$$
(2n+1) \in \{11,\ 13,\ 23,\ 25,\ 35,\ 37,\ \ldots\ldots\}
\tag{2.65}
$$

Case of Unbalanced distribution Let us consider the case of a linear distribution with $m = 3$ as shown in Fig. 2.8.

Fig. 2.8 Principle of the
back-EMF induction in a
distributed coil (linear
distribution)

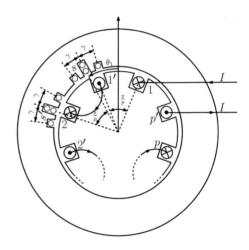

Table 2.2 K_{d1} of the balanced and linear distributions for $m = 3$

	$q = 1$	$q = 3$	$q = 5$	$q = 7$
Balanced distribution	0.6667	0.9598	0.9854	0.9926
Linear distribution	0.7500	0.9698	0.9891	0.9944

The resulting back-EMFs are expressed as follows:

$$
\begin{cases}
e_{s(1-1')} = \frac{N_c}{4} LD\Omega B_1 \cos\left(p(\theta_1 - \gamma) - \omega t\right) \\
e_{s(2-2')} = \frac{N_c}{2} LD\Omega B_1 \cos\left(p\theta_1 - \omega t\right) \\
e_{s(3-3')} = \frac{N_c}{4} LD\Omega B_1 \cos\left(p(\theta_1 + \gamma) - \omega t\right)
\end{cases}
\tag{2.66}
$$

Then, the total back-EMF E_d is expressed as:

$$
e_d = \frac{(2 + 2\cos p\gamma)}{4} N_s LD\Omega B_1 \cos\left(p\theta_1 - \omega t\right)
\tag{2.67}
$$

leading to the following winding factor of the fundamental back-EMF:

$$
K_{d1} = \frac{1 + \cos p\gamma}{2}
\tag{2.68}
$$

which is higher than the one given by the balanced distribution, as reported in
Table 2.2.

2.3.3   Root-Mean-Square of the Induced Back-EMF

2.3.3.1   Case of a Concentrated Winding

Concentrated windings are characterized by unity distribution factors of the fundamental and harmonic back-EMFs. Then the root-mean-square (rms) of the total induced back-EMF is expressed as:

$$E_{rms} = pN_cLD\Omega B_{rms} \tag{2.69}$$

where

$$B_{rms} = \sqrt{\frac{1}{\tau_p} \int_{\tau_p} B^2(\theta)d\theta} \tag{2.70}$$

which can be obtained using the spatial repartition of the flux density $B(\theta)$. Otherwise, it could be expressed in terms of the flux per pole Φ, such as:

$$\Phi = S_p \overline{B} \tag{2.71}$$

where S_p is the surface occupied by a pole within the air gap, such that:

$$S_p = \frac{\pi DL}{2p} \tag{2.72}$$

and where $\overline{B}$ is the average value of $B(\theta)$ under a pole, with:

$$\overline{B} = \frac{1}{\tau_p} \int_{\tau_p} B(\theta)d\theta \tag{2.73}$$

which is linked to B_{rms} by the form factor K_f as:

$$B_{rms} = K_f \overline{B} \tag{2.74}$$

Finally

$$E_{rms} = K_f pN_cLD\Omega\Phi\frac{2p}{\pi DL} = 4pK_fN_cf\Phi \tag{2.75}$$

where f is the frequency of the fundamental back-EMF.

2.3.4   Case of a Distributed Winding

The fundamental and the harmonics of the back-EMF have different distribution factors, then the root-mean-square (rms) of the total induced back-EMF could not be obtained using Eq. (2.69). Its formulation turns to be as follows:

$$E_{rms} = pN_c LD\Omega \sqrt{\frac{(K_{d1}B_1)^2}{2} + \frac{(K_{d2}B_2)^2}{2} + \dots\dots + \frac{(K_{d(2n+1)}B_{(2n+1)})^2}{2}}$$

(2.76)

which requires the knowledge of the harmonic content of the *Fourier* expansion of $B(\theta)$ and the distribution factor of each harmonic.

2.4   Developed Electromagnetic Torque

Fundamentally, the origin of the electromagnetic torque developed by AC machines is associated to two phenomena that could exist separately or conjointly, such that:

- the synchronization of two rotating fields: one created by a polyphase winding in the stator and the other is produced by a single or polyphase winding or permanent magnets (PMs) in the rotor, resulting in the so-called "synchronizing torque". This phenomenon concerns smooth pole AC machines, such as the induction and smooth pole synchronous machines where the air gap thickness is supposed quasi-constant with the slotting effect neglected.

  Moreover, machines equipped with PMs mounted on the rotor surface are classified in the smooth pole category is so far as the magnet permeability is almost equal to the air one. Actually, these machines are characterized by:

  - a mechanical air gap between the PM outer surface and the stator teeth, and
  - a magnetic air gap between the rotor yoke and the stator teeth.

  This latter is the one involved in the machine torque production,

- the effect of a rotating field created by a polyphase winding in the stator on a switched reluctance at the rotor surface, resulting in the so-called "reluctant torque". The corresponding AC machines are characterized by a magnetic circuit presenting a high anisotropy with remarkably-disproportioned values of the air gap width. It seems, at a first glance, that switched reluctance machines belong to such a class. However, with their armature fed by unipolar currents, they are far from producing a rotating field which makes it unfair their categorisation among the machines developing a reluctant torque.

  With this said, it should be underlined that the development of a reluctant torque is practically-feasible and is commonly allied to the one of a synchronizing torque. Indeed, salient pole field-excited synchronous, claw pole, interior PM-excited synchronous including spoke type, and other machines develop both synchronizing and reluctant torques.

The formulation of the synchronizing torque is derived hereunder.

2.4.1 Torque Production in Smooth Pole AC Machines

2.4.2 Principle

Let us consider, for instance, the principle od operation of a turbo-alternator, that is a smooth pole synchronous machine operating as a generator. Such operation is initiated by feeding the field by a DC current and driving the rotor at a speed Ω. Consequently a rotating field at the same speed is created in the air gap. This field causes variable fluxes crossing the phases of the armature in the stator and then the induction in these circuits of three phase back-EMFs with an angular frequency:

$$\omega = p_r \Omega \tag{2.77}$$

where p_r is the rotor pole pairs.

The connection of the armature to three phase balanced loads or to the grid leads to the circulation in its phases of three phase currents with an angular frequency ω in the stator circuits. A far as these circuits are shifted by a 120° electrical angle, they generate a second rotating field in the air gap moving at the speed:

$$\Omega' = \frac{\omega}{p_s} = \frac{p_r}{p_s} \Omega \tag{2.78}$$

where p_s is the stator pole pair.

Giving that $p_s = p_r = p$, the rotating field created by the armature (also known as the armature magnetic reaction) and the field one are then moving in synchronism. The condition of torque production is then satisfied, which makes possible the electromechanical conversion within the turbo-alternator. Consequently, an electric power is provided by the synchronous machine to the loads directly or through the grid.

2.4.3 Formulation

Let us call B_a and B_f the rotating fields created by the armature and the field, respectively, and let us assume that they are sinusoidally-shaped, with:

$$\begin{cases} b_a = B_a \cos{(p_s\theta - \omega t)} & = B_a \cos{\left(\frac{2p_s}{D}x - \omega t\right)} \\ b_f = B_f \cos{(p_r(\theta - \Omega t - \theta_0))} = B_f \cos{\left(\frac{2p_r}{D}x - p_r\Omega t - p_r\theta_0\right)} \end{cases} \quad (2.79)$$

where θ_0 is the mechanical shift between to two rotating waves at the initial time.

Accounting for the assumption that considers $(\mu_{Fe} \gg \mu_0)$, the magnetic energy W_{mag} turns to be concentrated in the air gap, and is expressed as:

$$W_{mag} = \int_V \frac{(b_a(x,t) + b_f(x,t))^2}{2\mu_0} dV \quad (2.80)$$

where V is the air gap volume.

The development of Eq. (2.80) gives:

$$W_{mag} = \frac{eL_s}{2\mu_0} \int_0^{\pi D} \left(b_a^2(x,t) + b_f^2(x,t) + 2b_a(x,t)b_f(x,t)\right) dx \quad (2.81)$$

The substitution of $b_a(x,t)$ and $b_f(x,t)$ by their expressions given in Eq. (2.79) yields:

$$W_{mag} = W_a + W_f + W_{af} \quad (2.82)$$

where:

$$\begin{cases} W_a = \frac{eL_s}{2\mu_0} \int\limits_0^{\pi D} b_a^2(x,t)dx & = \frac{eL_s}{2\mu_0} \frac{\pi D B_a^2}{2} \\ W_f = \frac{eL_s}{2\mu_0} \int\limits_0^{\pi D} b_f^2(x,t)dx & = \frac{eL_s}{2\mu_0} \frac{\pi D B_f^2}{2} \\ W_{af} = \frac{eL_s}{2\mu_0} \int\limits_0^{\pi D} 2b_a(x,t)b_f(x,t)dx = W_{af1} + W_{af2} \end{cases} \quad (2.83)$$

where:

$$\begin{cases} W_{af1} = \frac{eL_s}{2\mu_0} B_a B_f \int\limits_0^{\pi D} \cos{\left(\frac{2(p_s + p_r)}{D}x - (\omega + p_r\Omega)t - p_r\theta_0\right)} dx \\ W_{af2} = \frac{eL_s}{2\mu_0} B_a B_f \int\limits_0^{\pi D} \cos{\left(\frac{2(p_s - p_r)}{D}x - (\omega - p_r\Omega)t + p_r\theta_0\right)} dx \end{cases} \quad (2.84)$$

The case where $p_r \neq p_s$ is characterized by $W_{af1} = W_{af2} = 0$, which gives $W_{af} = 0$, and:

$$W_{mag} = W_a + W_f = \frac{eL_s\pi D}{2\mu_0} \left(\frac{B_a^2 + B_f^2}{2}\right) \quad (2.85)$$

However, if $p_r = p_s = p$, then $W_{af1} = 0$ and:

$$W_{af} = W_{af2} = \frac{eL_s\pi DB_aB_f}{2\mu_0} \cos\left(-(\omega - p_r\Omega)t + p_r\theta_0\right) \qquad (2.86)$$

which leads to:

$$W_{mag} = W_a + W_f + W_{af2} = \frac{eL_s\pi D}{2\mu_0}\left(\frac{B_a^2 + B_f^2}{2} + B_aB_f\cos\left(-(\omega - p\Omega)t + p\theta_0\right)\right) (2.87)$$

Assuming that the magnetic circuit is not saturated, the electromagnetic torque T_{em} is expressed as follows:

$$T_{em} = \frac{\partial W_{mag}}{\partial\theta_0} \qquad (2.88)$$

Thus, if $p_r \neq p_s$ then:

$$T_{em} = 0 \qquad (2.89)$$

However, if $p_r = p_s = p$, then:

$$T_{em} = p\frac{eL_s\pi D}{2\mu_0}B_aB_f\sin\left((\omega - p\Omega)t - p\theta_0\right) \qquad (2.90)$$

which has a null average value except for:

$$\Omega = \frac{\omega}{p} \qquad (2.91)$$

that is to say the speed of the rotating field created by the armature $\frac{\omega}{p}$ is equal to the speed of the rotating field created by the field, which is called the synchronism condition. In this case, the expression of the electromagnetic torque is limited to:

$$T_{em} = -p\frac{eL_s\pi D}{2\mu_0}B_aB_f\sin\left(p\theta_0\right) \qquad (2.92)$$

Accounting for the relations between the amplitudes of the flux densities and the MMFs of the armature and the field, such that:

$$\begin{cases} B_a = \frac{\mu_0}{e}F_a \\ B_f = \frac{\mu_0}{e}F_f \end{cases} \qquad (2.93)$$

the expression of the electromagnetic torque given by Eq. (2.92) turns to be:

$$T_{em} = -p \frac{\mu_0 L_s \pi D}{2e} F_a F_f \sin(p\theta_0) \tag{2.94}$$

Let us call Λ the air gap permeance which is defined as:

$$\Lambda = \mu_0 \frac{L_s D}{e} \tag{2.95}$$

then Eq. (2.94) is reduced to:

$$T_{em} = -\frac{\pi}{2} p \Lambda F_a F_f \sin(p\theta_0) \tag{2.96}$$

2.5 Conclusion

The second chapter treated, in a first part, the formulation of the rotating fields on which lies the principle of operation of AC machines. This has been achieved considering:

(i) the case of a single phase winding linked to the rotor and fed by a DC current,
(ii) the cases of different poly-phase windings fed by poly-phase sinusoidal currents that could be linked either to the stator or the rotor.

Then, the effects of the generated rotating fields on the windings located in the other side of the air gap, are analyzed. These effects are initiated by the induction of back-EMFs in the windings. A special attention has been paid to their formulation considering concentrated and distributed windings. These later are characterized by a distribution factor which has been formulated considering the fundamental as well as the harmonics of the back-EMF.

Finally, the electromagnetic torque production of AC machines has been focused. To start with, the phenomena on which are based the torque production have been analyzed. Then, the electromagnetic torque formulation has been carried out in the case of smooth pole AC machines; the case of salient pole ones will be developed within a course dedicated to emergent AC machines.

It has been found that one of the fundamental conditions that enable the torque production is the similitude of the polarities in both sides of the air gap. However, in recent years, it has been shown that such a condition is no longer mandatory, providing that the armature is equipped by the so-called "fractional-slot concentrated windings". These are the subject of the following chapter.

Chapter 3
Fractional-Slot Concentrated Windings: Design and Analysis

Abstract The arrangement of either distributed or concentrated windings in slots around the air gap of AC machines represents a crucial design issue, in so far as it deeply affects their features. This statement is much more critical in fractional-slot concentrated winding permanent magnet synchronous machines, also named fractional-slot permanent magnet machines (FSPMMs). These are characterized by fractional slot per pole and per phase lower than unity, resulting in dense harmonic contents of the armature MMF with a number of pole pairs lower than the field one. This chapter is aimed at the arrangement of the armature winding of FSPMMs, which is achieved considering the star of slots approach. Following the winding arrangement, the star of slots is used for the investigation of the back-EMF harmonic content as well as for the determination of the winding factors of the fundamental and harmonic backs-EMFs. Finally a case study is treated with emphasis on the MMF spatial repartition and harmonic content.

Keywords Fractional-slot concentrated winding
Permanent magnet synchronous machines · Star of slots approach
Double layer slots · Single layer slots · Back-EMF · MFF spatial repartition
Fourier expansion

3.1 Introduction

In recent years, a new area of electric machine technology has evolved, based on the principle that the best machine design is that which produces the optimum-match between the machine and the power electronic converter, yielding the so-called "converter fed machines". These are presently given an increasing attention where many conventional machine design considerations are rethought, such as the number of phases, the number of pole pairs, the magnetic circuit material and geometry, the flux paths (radial, axial, or transversal), the winding shape and distribution, and so on.

Of particular interest is the substitution of distributed windings by concentrated ones. In recent years, concentrated windings are mainly integrated in brushless

© The Author(s) 2019
A. Masmoudi, *Design and Electromagnetic Feature Analysis*
of AC Rotating Machines, SpringerBriefs in Electrical and Computer
Engineering, https://doi.org/10.1007/978-981-13-0920-5_3

machines, with their numbers of slot per pole and per phase fractional lower than unity, leading to the so-called "fractional slot concentrated winding permanent magnet machines" or simply "concentrated winding permanent magnet machines" (FSP-MMs).

An increasing attention is presently given to FSPMMs and their integration in different applications covering a wide power range starting from domestic equipments to wind generators, going through electric and hybrid vehicles. The growing interest in FSPMMs is motivated by [1, 2]:

- their low copper losses achieved thanks to their short end-windings,
- their reduced cogging torque ripple,
- their high fault tolerance capability,
- their wide flux weakening range.

With this said, it has reported that the above-listed attractive features are achieved for appropriate slot/pole combinations. Otherwise, FSPMMs would be penalized by many drawbacks such as: a low winding factor leading to reduced torque production capability, a high cogging torque, a high iron losses especially in the permanent magnets, and unbalanced magnetic forces leading to a high vibration and noise.

Fractional-slot concentrated windings are far from being a new concept in machine design. They have been reported in a patent published in 1895 [3]. However, they have been unexploited during almost a century characterized by the supremacy of distributed windings. Starting from 2002, They have been integrated in high power three-phase synchronous machines. Since then, they turn to be very trendy.

In [4], Cros and Viarouge presented an enlightening study dealing with the integration of concentrated windings in high performance brushless machines. They identified the various slot/pole combinations that can support three-phase concentrated windings. Furthermore, they presented a systematic method to find out the optimum concentrated winding layout in both cases of regular and irregular slot distribution. They proposed some guidelines to identify the slot/pole combinations that can provide high performance designs. They presented and discussed the features exhibited by sample designs using concentrated windings. They demonstrated that these design offer improved features compared to those of distributed winding ones with a unity slot per pole and per phase. Finally, they proposed a technique to improve the machine manufacturability by using segmented soft magnetic composites.

In [5], EL-Refaie et al. expanded the work presented in [4] to cover 4, 5, and 6 phase FSPMMs. Tables including the winding factors, cogging torque indicators, and net radial force indicators for different slot/pole combinations have been provided. Moreover, a key parameter for the selection of the optimal slot/pole combination has been introduced: the rotor loss figure of merit (FOM) that enables the comparison of the rotor iron losses for the various slot/pole combinations on relative basis. Values of the rotor loss FOM have been evaluated for 3, 4, 5 and 6 phase FSPMMs.

In [6], Libert and Soulard investigated various slot/pole combinations for surface-mounted PM machines equipped with double-layer fractional-slot concentrated windings. Among the considered factors were the winding factors, the MMF harmonic content, the torque ripple, and the radial magnetic forces. To do so, they considered the approach developed in [4]. They have found that machine candidates characterized by asymmetrical winding layouts have unbalanced radial magnetic forces causing vibration and noise within the stator.

In [7, 8], Bianchi et al. presented another approach to determine the optimal winding layout for various slot/pole combinations of FSPMMs based on the star of slots. This approach has been introduced since the fifteens [9, 10], to design large alternators including a high number of poles. It has been found a useful tool to find out the suitable coil connections that maximize the fundamental (synchronous) component of the back-EMF induced in the armature. In recent years, it has been updated in order to suitably arrange concentrated windings equipping FSPMMs. It has been reported in [11], that the star of slots approach enables a suitable arrangement of PM machines equipped fractional-slot concentrated windings inserted in double- or single-layer slots with an investigation of the harmonic contents of their back-EMFs and armature MMFs.

In [12], Magnussen and Sadarangani carried out a comparative study between distributed windings with a unity slot per pole and per phase, single-layer and double-layer fractional-slot concentrated windings. They have found that by choosing appropriate slot/pole combinations, FSPMMs exhibit lower copper losses and cogging torque than those of distributed windings. Furthermore, it has been found that fractional-slot concentrated windings inserted in double-layer slots have the shortest axial length and hence exhibits the best compactness among the three arrangements under comparison.

In [13], Zhu summarized the various structural and design features of FSPMMs for EV/HEV applications. He has reported that FSPMMs have already been employed in the commercial HEVs thanks to their high torque density, high efficiency, low torque ripple, good flux-weakening and fault-tolerance performance. However, due to the dense harmonic content of their armature MMF, FSPMMs exhibit relatively high rotor eddy current loss, potentially high unbalanced magnetic force and acoustic noise and vibration, while the reluctance torque component is relatively low or even negligible when an interior PM rotor is employed.

In [14], Ben Hamadou et al. presented a new integrated electromechanical set for automotive hybrid propulsion systems. It consists in two concentric permanent-magnet(PM) machines with inner and outer surface-mounted PM rotors, and a stator sandwiched between the two rotors, yielding a so-called "single-stator dual-rotor PM machine". The stator magnetic circuit consists of two concentric parts decoupled by a nonmagnetic ring. They are equipped by fractional-slot concentrated windings to extend the flux weakening range. A special attention has been paid to the formulation and the analysis of the armature and PM MMF harmonic contents with an optimization of two influent circumferential parameters based on the minimization of the MMF total harmonic distortion. The torque production capability of the optimized concept has been investigated by FEA.

In [15], Abdennadher and Masmoudi developed a star of slots-based approach to design a class of FSPMMs where the armature is arranged by connecting in parallel the coils or suitable combinations of coils of each phase. Accordingly, the integration of these machines in 48 V technology based automotive systems turns to be realistic. Moreover, the adopted approach enabled a significant improvement of the fault-tolerance capability under open-circuit faulty operation. The proposed approach has been initiated by an identification allied to a characterization of the slot/pole combinations enabling the parallel connection of the phase coils or group of coils. Then, the features of selected three and five phase FSPMMs have been investigated by a 2D FEA, under healthy operation, and under different open-circuit faulty scenarios, with emphasis on their torque production capabilities.

3.2 Concentrated Winding Arrangement Based on the Star of Slots Approach

3.2.1 Study Background

As previously introduced, the star of slots approach has been established since the fifteens [9, 10], to design large alternators including a high number of poles. Basically, suitable coil distributions of double-layer windings, maximizing the synchronous component of the back-EMF, have been achieved thanks to the application of the star of slots approach. Fundamentally and for given speed and torque, an increase of the back-EMF leads to a decrease of the armature current, and consequently an improvement of the efficiency.

Recently, the star of slot approach has been reconsidered, by several teams [2, 11–14], for the arrangement of concentrated windings equipping FSPMMs. Many potentialities of the approach have been established, such as:

- the arrangement of double-layer fractional-slot concentrated windings,
- the extension to the case of single-layer fractional-slot concentrated windings,
- the analysis of the harmonic content of the back-EMF waveform,
- the analysis of the harmonic content of air-gap MMF spatial repartition.
- the graphical determination of the winding factors of the fundamental and harmonics of the back-EMF.

3.2.2 Basis of the Star of Slots

The principle of the star of slots consists in distributing the coils of the armature phases taking into account the number of stator slots and the one of the rotor poles.

Assuming that in all machine coils are induced sinusoidally-shaped back-EMFs which are shifted according to the electrical positions of the corresponding coils with respect to the PM-rotating field. Consequently, the back-EMFs can be expressed in terms of complex quantities and represented by space vectors.

From a geometrical point of view, the star of slots is simply reduced to a phasor diagram made up of the space phasors representing the back-EMFs induced in the

Fig. 3.1 Numeration of the slots filled by a double-layer concentrated winding

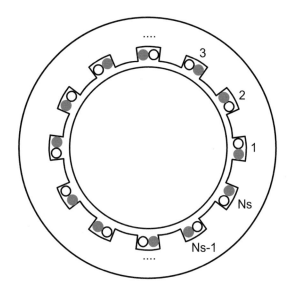

machine coils. In what follows, the methodology of drawing the star of slots in the case of double-layer windings is described. Then, its effectiveness in distributing the coils through the machine phases is highlighted. Finally, its extension to the case of single-layer windings is treated.

3.2.3 Arrangement of Double-Layer Concentrated Windings

3.2.3.1 Considerations and Objective

Let us consider a double-layer concentrated winding inserted in N_s stator slots. Therefore, the armature phases include N_s coils which are numbered consequently as illustrated in Fig. 3.1.

In the manner of the slots, the phasors representing the back-EMFs induced in the N_s coils are numbered using the same numerals and taking into account the contribution of just one layer among the two of each coil (for instance the one filling the slot left-side). Obviously, the back-EMF induced in a coil should be represented by a phasor equal to the vectorial sum of the ones corresponding to the back-EMFs induced in its two layers. These two phasors have the same modulus and speed but are shifted by an electrical angle corresponding to the coil pitch.

With this said, accounting for just one layer of the coil in the representation of the back-EMF phasor does absolutely not affect the correctness of the star of slots approach. The objective of this later could be pointed up by the following key rules:

- the armature phases should be shifted by $\frac{2\pi}{q}$-electrical angle,

- the back-EMFs induced in the armature phases should make up a balanced q-phase voltage system,
- the amplitude of the main harmonic of the back-EMF should be maximum.

3.2.3.2 Methodology

Let us call δ the greatest common divisor (GCD) of the number of slots N_s and the number of pole pairs p. It represents the machine periodicity and corresponds to the number of identical parts in which the machine could be split. That is to say that the star of slots is made up of $\frac{N_s}{\delta}$ spokes each one includes δ back-EMF phasors.

From a geometrical point of view, this would lead to a star of slots including N_s phasors equally displaced in the case of a unity value of δ. However, if δ is greater than unity, one would find after drawing the first $\frac{N_s}{\delta}$ phasors that:

- the phasor numbered $\frac{N_s}{\delta} + 1$ would be aligned with the one numbered 1,
- the phasor numbered $\frac{N_s}{\delta} + 2$ would be aligned with the one numbered 2,
- until the phasor numbered $\frac{2N_s}{\delta}$ would be aligned with the one numbered $\frac{N_s}{\delta}$.

This procedure is repeated until δ turns are achieved to draw the N_s back-EMF phasors.

Giving the fact that the N_s slots are equidistant, the electric angular displacement α_e between the phasors induced in the coils of two adjacent slots is expressed as follows:

$$\alpha_e = p \left(\frac{2\pi}{N_s} \right) \tag{3.1}$$

Thus, one can draw the phasors successively taking into account angle α_e.

Following the representation of the N_s phasors, the next step consists in the identification of the phasors of each phase (that is the slots allocated to each phase). To do so, the following graphical procedure is carried out:

- two opposite sectors with an angular opening of $\alpha_s = \frac{\pi}{q}$ each are delimited. The back-EMF phasors located inside the two sectors belong to the same phase.
- a positive polarity is assigned to the coils with the corresponding back-EMF phasors within a sector, while a negative polarity is assigned to those with the corresponding back-EMF phasors are located in the opposite sector,
- the two opposite sectors shifted by an angle $\frac{2\pi}{q}$ include the back-EMF phasors of the next phase. Repeating this step q times allows the identification of the slots allocated to the q phases.

It should be noted that the two opposite sectors assigned to a given phase would not necessary include the same number of phasors. Moreover, some slot/pole combination yield the totality of the back-EMF phasors located in one sector while the opposite one is vacant, such as the case of $N_s = 9$, $p = 3$ and $q = 3$, to be treated in the following paragraph.

3.2.3.3 Case of a Unity GCD

Case of an Even Number of Spokes Let us consider a double-layer fractional-slot concentrated winding characterized by $N_s = 12$, $p = 5$, and $q = 3$, which leads to:

- a slot per pole and per phase $m = \frac{2}{5}$,
- a periodicity factor $\delta = 1$,
- an electrical shift between the back-EMFs of the coils of adjacent slots $\alpha_e = \frac{5\pi}{6}$.

Thus, the star of slots includes twelve spokes geometrically shifted by $\alpha = \frac{\pi}{6}$, and including one back-EMF phasor each.

Figure 3.2 illustrates three steps of drawing the star of slots, with in:

(a) are drawn the three first phasors,
(b) are drawn the twelve phasors,
(c) are delimited the two opposite sectors of each phase, with two phasors per sector.

An arrangement of the armature phases according to the complete star of slots shown in Fig. 3.2c, is illustrated in Fig. 3.3.

Case of an Odd Number of Spokes Let us consider a double-layer concentrated winding characterized by $N_s = 9$, $p = 4$ and $q = 3$, which gives $m = \frac{3}{8}$, $\delta = 1$ and $\alpha_e = \frac{8\pi}{9}$. Therefore, the star of slots includes nine spokes geometrically shifted by $\alpha = \frac{2\pi}{9}$, and including one back-EMF phasor each.

Figure 3.4 illustrates three steps of drawing the star of slots, with in:

(a) are drawn the three first phasors,
(b) are drawn the nine phasors,
(c) are delimited the two opposite sectors of each phase.

Referring to Fig. 3.4c, it is to be noted that in the case of odd number of spokes, the two opposite sectors assigned to a phase do not include the same number of back-EMF phasors.

3.2.3.4 Case of a GCD Greater Than Unity

Case of an Even Number of Spokes Let us consider a double-layer concentrated winding characterized by $N_s = 12$, $p = 10$ and $q = 3$, that gives $m = \frac{1}{5}$, $\delta = 2$ and $\alpha_e = \frac{5\pi}{3}$. The resulting star of slots includes six spokes geometrically shifted by $\alpha = \frac{\pi}{3}$, and including two back-EMF phasors each.

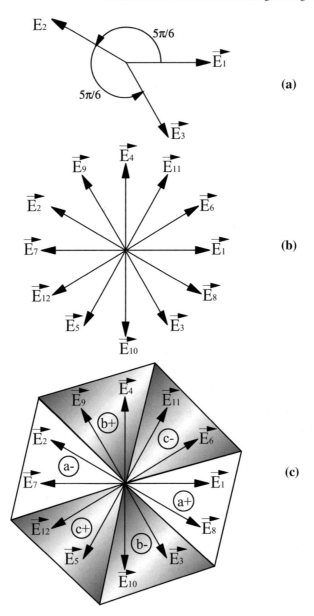

Fig. 3.2 Star of slots and phase coil identification of a double-layer concentrated winding characterized by $N_s = 12$, $p = 5$, $q = 3$, $\delta = 1$, a number of spokes equal 12, and $\alpha_e = \frac{5\pi}{6}$

Fig. 3.3 Armature winding
arrangement of a
fractional-slot concentrated
winding surface mounted
PM synchronous machine
characterized by $N_s = 12$,
$p = 5$, and $q = 3$

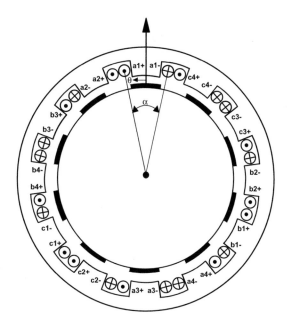

Figure 3.5 illustrates three steps of drawing the star of slots, with in:

(a) are drawn the two first phasors,
(b) are drawn the twelve phasors,
(c) are delimited the opposite sectors of each phase, with two aligned phasors per
    sector.

Figure 3.5c highlights the fact that δ represents definitely the periodicity of the
winding. In fact, the same star of slots could be obtained considering two identical
double-layer concentrated windings characterized by $N_s = 6$, $p = 5$ and $q = 3$.

Case of an Odd Number of Spokes Let us consider a double-layer concentrated
winding characterized by $N_s = 9$, $p = 3$ and $q = 3$, that gives $m = \frac{1}{2}$, $\delta = 3$ and
$\alpha_e = \frac{2\pi}{3}$. Consequently, the star of slots includes three spokes geometrically shifted
by $\alpha = \frac{2\pi}{3}$, and including three back-EMF phasors each.

Figure 3.6 illustrates three steps of drawing the star of slots, with in:

(a) are drawn the three first phasors,
(b) are drawn the nine phasors,
(c) are delimited the two opposite sectors of each phase.

Once more, Fig. 3.6c confirms the fact that in the case of odd number of spokes,
the two opposite sectors assigned to a phase do not include the same number of
phasors. Moreover, in the case where $\alpha_e = \alpha$, the totality of the back-EMF phasors
of each phase are located in one sector while the opposite one is vacant. This offers
the possibility to connect:

Fig. 3.4 Star of slots and
phase coil identification of a
double-layer concentrated
winding characterized by
$N_s = 9$, $p = 4$, $q = 3$,
$\delta = 1$, a number of spokes
equal 9, and $\alpha_e = \frac{8\pi}{9}$

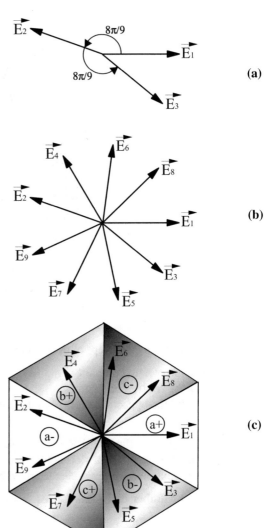

- the phase coils in series in which case an open-circuit failure of one of them would affect the totality of the phase,
- the phase coils in parallel in which case an open-circuit failure of one of them would affect just one branch among the $\frac{N_s}{q}$ parallel ones. However, the designer should be aware of the circulating currents due to the harmonic back-EMFs. These could be a serious limitation of the parallel connection of the phase coils.

Fig. 3.5 Star of slots and phase coil identification of a double-layer concentrated winding characterized by $N_s = 12$, $p = 3$, $q = 3$, $\delta = 2$, a number of spokes equal 6, and $\alpha_e = \frac{5\pi}{3}$

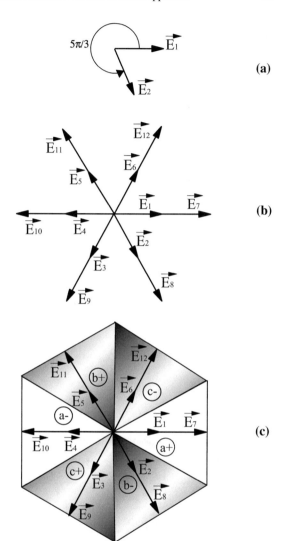

(a)

(b)

(c)

3.2.3.5 Particular Cases

The previously stated rules could not be applied to any case. This paragraph gives an example of a double-layer concentrated winding characterized by $N_s = 12$, $p = 5$ and $q = 6$, which leads to $\delta = 1$ and $\alpha_e = \frac{5\pi}{3}$. The resulting star of slot includes twelve spokes geometrically shifted by $\alpha = \frac{\pi}{6}$, and including one back-EMF phasor each, which is the same as the one illustrated in Fig. 3.2b.

However, as far as the number of phases is equal to 6 instead of 3, the two sectors, delimiting the back-EMF phasors that belong to a phase, have an angular opening of

Fig. 3.6 Star of slots and phase coil identification of a double-layer concentrated winding characterized by $N_s = 9$, $p = 3$, $q = 3$, $\delta = 3$, a number of spokes equal 3, and $\alpha_e = \frac{2\pi}{3}$

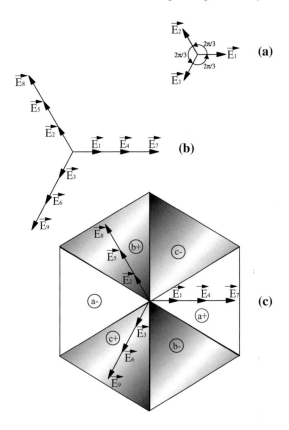

Table 3.1 The back-EMF phasors of the six phases with the corresponding polarities in the case of an angular shift between the two sectors assigned to a given phase equal to π

Phase	Back-EMF phasors	Phase	Back-EMF phasors
"a"	$\vec{E}_1^+$ and $\vec{E}_7^-$	"d"	$\vec{E}_7^+$ and $\vec{E}_1^-$
"b"	$\vec{E}_{11}^+$ and $\vec{E}_5^-$	"e"	$\vec{E}_5^+$ and $\vec{E}_{11}^-$
"c"	$\vec{E}_9^+$ and $\vec{E}_3^-$	"f"	$\vec{E}_3^+$ and $\vec{E}_9^-$

$\frac{\pi}{6}$ each. Moreover, the sectors of a phase are shifted from those of the previous one by $\frac{\pi}{3}$.

If the two sectors assigned to a given phase would be opposite as previously considered, the resulting repartition would be the one classified in Table 3.1.

One can notice that while even slots are not involved in the winding repartition, the odd ones are occupied by two phases with opposite polarization, which is practically unfeasible. This problem is solved considering an angular shift between the two sectors assigned to a given phase equal to $\pi + \alpha$, leading to the repartition classified in Table 3.2.

Table 3.2 The back-EMF phasors of the six phases with the corresponding polarities in the case of an angular shift between the two sectors assigned to a phase equal to $\pi + \alpha$

Phase	Back-EMF phasors	Phase	Back-EMF phasors
"a"	$\overrightarrow{E}_1^+$ and $\overrightarrow{E}_{12}^-$	"d"	$\overrightarrow{E}_7^+$ and $\overrightarrow{E}_6^-$
"b"	$\overrightarrow{E}_{11}^+$ and $\overrightarrow{E}_{10}^-$	"e"	$\overrightarrow{E}_5^+$ and $\overrightarrow{E}_4^-$
"c"	$\overrightarrow{E}_9^+$ and $\overrightarrow{E}_8^-$	"f"	$\overrightarrow{E}_3^+$ and $\overrightarrow{E}_2^-$.

The corresponding star of slots including the sectors assigned to each of the six phases, as well as the machine cross-section are shown in Fig. 3.7.

3.2.4 Arrangement of Single-Layer Concentrated Windings

Basically, the star of slots has been introduced to suitably arrange double-layer concentrated windings, in an attempt to maximize the synchronous component of the back-EMF, and consequently reduce its harmonic content.

In recent years, an approach to suitably deduce the arrangements of single-layer concentrated windings from those of double-layer ones, sharing the same values of N_s, p and q, has been reported in the literature [8, 11]. It simply consists in removing the back-EMF phasors with even numbers in the star of slots of the double-layer windings. The resulting star of slots would include just the back-EMF phasors of the coil-sides located in the slots with odd numbers; consideration already agreed in the case of double-layer windings.

The application of the above-described approach to find out the single-layer windings corresponding to the previously-studied double-layer ones, characterized by an even number of slots, is illustrated in:

- Figure 3.8 illustrating the transformation of the star of slots from a double layer distribution to a feasible single layer of a concentrated winding characterized by $N_s = 12$, $p = 5$, and $q = 3$,
- Figure 3.9 illustrating the transformation of the star of slots from a double layer distribution to a feasible single layer of a concentrated winding characterized by $N_s = 12$, $p = 10$, and $q = 3$.

One can notice that the obtained single-layer concentrated windings are balanced. However, removing the back-EMF phasors with even numbers in the star of slots of the double-layer windings of machines characterized by an odd number of slots leads to unbalanced and therefore unfeasible single-layer distributions, such those shown in:

- Figure 3.10 illustrating the transformation of the star of slots from a double layer distribution to an unfeasible single layer of a concentrated winding characterized by $N_s = 9$, $p = 4$ and $q = 3$,

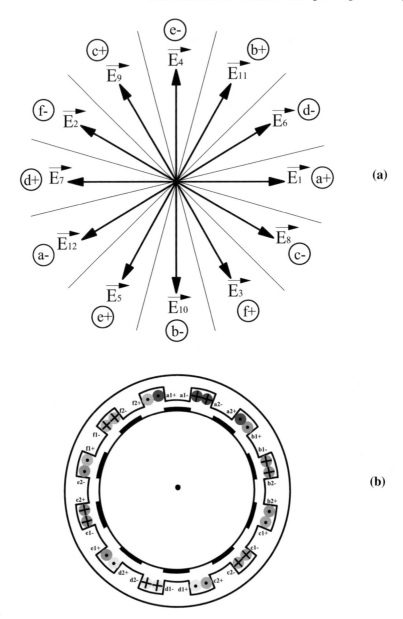

(a)

(b)

Fig. 3.7 Repartition of a double-layer concentrated winding for which $N_s = 12$, $p = 5$, $q = 6$, and $\delta = 1$, with a number of spokes equal to one, and an angular shift $\alpha_e = \frac{5\pi}{3}$. Legend: (**a**) star of slots, (**b**) machine cross-section

Fig. 3.8 Transformation of the star of slots of a double-layer concentrated winding leading to feasible single-layer distribution, characterized by $N_s = 12$, $p = 5$, and $q = 3$

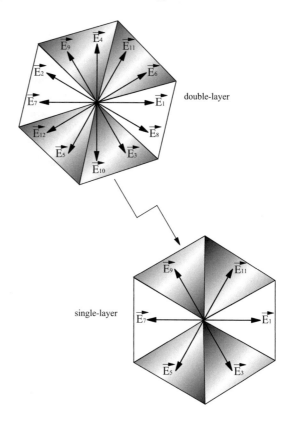

double-layer

single-layer

- Figure 3.11 illustrating the transformation of the star of slots from a double layer distribution to an unfeasible single layer of a concentrated winding characterized by $N_s = 9$, $p = 3$ and $q = 3$.

Furthermore, in the case of odd number of slots, it would be impossible to reach a single-layer distribution as far as the coil numbered N_s would necessary share the slot number one with the coil numbered one.

3.3 Winding Factor Star of Slots Based-Prediction

Generally speaking, The winding factor K_w is expressed as follows:

$$K_w = K_d K_r K_s \tag{3.2}$$

where:

- K_d is the distribution factor,

Fig. 3.9 Transformation of
the star of slots of a
double-layer concentrated
winding leading to feasible
single-layer distribution,
characterized by $N_s = 12$,
$p = 3$, and $q = 3$

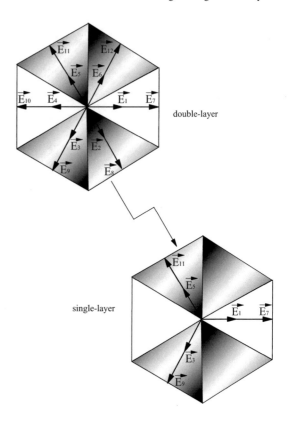

- K_r is the factor that accounts for the reduced coil pitch with respect to the pole
  pitch,
- K_s is the skewing factor. The skewing of the slots is used to eradicate the torque
  ripple due to slotting or the cogging torque in case of PM synchronous machines.
  As far as FSPMMs exhibit low cogging torque, the slot skewing turns to be useless
  and $K_s = 1$.

   In what follows, the star of slots is used to find out the winding factor of the
fundamental back-EMF in both cases of double and single layer slots.

3.3.1 *Winding Factor of the Fundamental Back-EMF*

3.3.1.1 Case of Double Layer Slots

Basically, the winding factor is the ratio of the back-EMF induced in the distributed
winding over the one of the winding assumed concentrated in just one coil. It takes
into account the the phase-shift between the back-EMFs induced in the different coils

Fig. 3.10 Transformation of
the star of slots of a
double-layer concentrated
winding leading to
unfeasible single-layer
distribution, characterized by
$N_s = 9$, $p = 4$, and $q = 3$

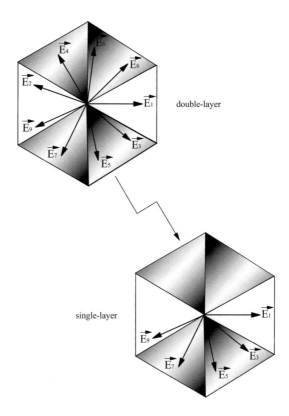

of the distributed winding (K_d) as well as the reduced coil pitch with respect to the
pole pitch (K_r).

Let us consider, for instance, the double layer winding that has the star of slots
illustrated in Fig. 3.2c which is recalled in Fig. 3.12a. One can draw the phasors of
the back-EMFs induced in the four coils of phase "a", as shown in Fig. 3.12b. The
vectorial sum of the back-EMFs is illustrated in Fig. 3.12c. The winding factor K_{wf}^{dl}
of the fundamental back-EMF could be expressed as follows:

$$K_{wf}^{dl} = \frac{4 \parallel \overrightarrow{E}_2 \parallel \left(1 + \cos\left(\frac{\pi}{6}\right)\right)}{8 \parallel \overrightarrow{E}_2 \parallel} = \frac{1 + \cos\left(\frac{\pi}{6}\right)}{2} \simeq 0.9330127 \qquad (3.3)$$

3.3.1.2   Case of Single Layer Slots

A graphical approach similar to the above-described one, applied to the star of slots
of the single layer winding, illustrated in Fig. 3.8, has enabled the prediction of the
winding factor K_{wf}^{sl} of the fundamental back-EMF, as follows:

Fig. 3.11 Transformation of
the star of slots of a
double-layer concentrated
winding leading to
unfeasible single-layer
distribution, characterized by
$N_s = 9$, $p = 3$, and $q = 3$

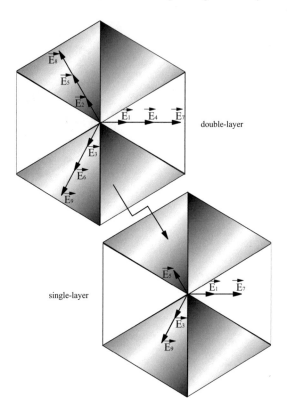

$$K_{wf}^{sl} = \frac{2 \left(2 \parallel \overrightarrow{E}_1 \parallel \sin \left(\frac{5\pi}{12} \right) \right)}{4 \parallel \overrightarrow{E}_1 \parallel} = \sin \left(\frac{5\pi}{12} \right) \simeq 0.9659258 \qquad (3.4)$$

3.3.2 Winding Factor of the Harmonic Back-EMFs

3.3.2.1 Graphical Methodology

Giving the fact that the armature pole pair ($p_s = 1$) is lower than the rotor one
($p_r = 5$), it is expected that the back-EMF induced in the armature presents sub-
and super-harmonics. In order to investigate the harmonic content of the back-EMF,
one should draw the star of slots for the different harmonic ranks including the
sub-harmonics.

   To do so, the back-EMF phasors $\overrightarrow{E}_1$, $\overrightarrow{E}_2$, ..., $\overrightarrow{E}_{N_s}$ should be drawn considering an angular shift α_k between two successive phasors, such that:

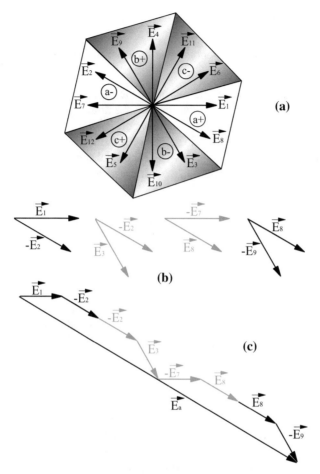

Fig. 3.12 Star of slots based-prediction of the winding factor of the fundamental back-EMF induced in the double-layer concentrated winding characterized by $N_s = 12$, $p = 5$, and $q = 3$

$$\alpha_k = k\frac{2\pi}{N_s} \qquad (3.5)$$

where k in the harmonic rank (Fig. 3.12).

Then, the back-EMF phasors of a given phase, identified by the star of slots corresponding to the fundamental back-EMF, are summed. If the vectorial sum is null, one can conclude that the considered harmonic is not present in the back-EMF spectrum. Otherwise, one can predict the winding factor of the corresponding harmonic back-EMF in the case of double layer slots. An approach, similar to the one adopted in the case of the fundamental back-EMF, is considered to find out the winding factor of a given harmonic back-EMF in the case of single layer slots.

3.3.2.2 Case Study

Let us consider the case of a double layer concentrated winding characterized by $N_s = 12$, $p = 5$, and $q = 3$.

Sub-Harmonic of Rank One The sub-harmonic of rank one has an angular shift between two successive $\alpha_1 = \frac{\pi}{6}$. This enabled the drawing of the star of slots corresponding to the considered sub-harmonic which is shown in Fig. 3.13a.

The coils of phase "a", identified in the star of slots illustrated in Fig. 3.2c, have the back-EMF phasors shown in Fig. 3.13b. Their vectorial sum, illustrated in Fig. 3.13c, allows the prediction of the winding factor K_{wf1}^{dl} of the subharmonic of rank 1, as follows:

$$K_{wf1}^{dl} = \frac{4 \parallel \overrightarrow{E}_2 \parallel \left(1 - \cos\left(\frac{\pi}{6}\right)\right)}{8 \parallel \overrightarrow{E}_2 \parallel} = \frac{1 - \cos\left(\frac{\pi}{6}\right)}{2} \simeq 0.066987 \qquad (3.6)$$

One can the low weight of the harmonic of rank one in the back-EMF of the double layer winding.

Discarding the even back-EMF phasors in the star of slots of Fig. 3.13a yields the star of slots of the sub-harmonic of rank one in the case of single layer slots. The vectorial sum of the back-EMF phasors of phase "a" (the coil located in the slot pair 1-2 and the one located in the slot pair 7-8) enables the prediction of the winding factor K_{wf1}^{sl}, as:

$$K_{wf1}^{sl} = \frac{4 \parallel \overrightarrow{E}_1 \parallel \sin\left(\frac{\pi}{12}\right)}{4 \parallel \overrightarrow{E}_1 \parallel} = \sin\left(\frac{\pi}{12}\right) \simeq 0.2588190 \qquad (3.7)$$

Sub-Harmonic of Rank Two The sub-harmonic of rank two has an angular shift between two successive $\alpha_2 = \frac{\pi}{3}$. This enabled the drawing of the star of slots corresponding to the considered sub-harmonic which is shown in Fig. 3.14a.

The coils of phase "a", identified in the star of slots illustrated in Fig. 3.2c, have the back-EMF phasors shown in Fig. 3.14b. One can clearly notice that the vectorial sum of these back-EMF phasors leads to $\overrightarrow{0}$. Consequently, the winding factor K_{wf2}^{dl} of the sub-harmonic of rank two is null.

Discarding the even back-EMF phasors in the star of slots of Fig. 3.14a yields the star of slots of the sub-harmonic of rank one in the case of single layer slots. The vectorial sum of the back-EMF phasors of phase "a" leads to $\overrightarrow{0}$, and then the winding factor $K_{wf2}^{sl} = 0$.

Sub-Harmonic of Rank Three The sub-harmonic of rank three has an angular shift between two successive $\alpha_3 = \frac{\pi}{2}$. This enabled the drawing of the star of slots corresponding to the considered sub-harmonic which is shown in Fig. 3.15a.

Fig. 3.13 Star of slots based-prediction of the winding factor of the sub-harmonic of rank one of the back-EMF induced in the double-layer concentrated winding characterized by $N_s = 12$, $p = 5$, and $q = 3$

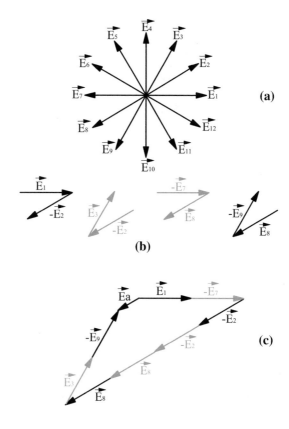

(a)

(b)

(c)

The coils of phase "a", identified in the star of slots illustrated in Fig. 3.2c, have the back-EMF phasors shown in Fig. 3.15b. The winding factor K^{dl}_{wf2} of the sub-harmonic of rank three is simply deduced from Fig. 3.15b, that is: $K^{dl}_{wf3} = 0.5$.

Discarding the even back-EMF phasors in the star of slots of Fig. 3.15a yields the star of slots of the sub-harmonic of rank three in the case of single layer slots. The vectorial sum of the back-EMF phasors of phase "a" enables the prediction of the winding factor $K^{sl}_{wf3} = 0.5$.

Super-Harmonic of Rank Seven The sub-harmonic of rank one has an angular shift between two successive $\alpha_7 = \frac{7\pi}{6}$. This enabled the drawing of the star of slots corresponding to the considered sub-harmonic which is shown in Fig. 3.16a. The coils of phase "a" have the back-EMF phasors shown in Fig. 3.16b with their vectorial sum illustrated in Fig. 3.16c. Although, the resulant back-EMF $\overrightarrow{E}_{a7}$ has a different phase-shift from the one of the fundamental back-EMF, it leads to the same winding factor $K^{dl}_{wf7} \simeq 0.9330127$. Similarly, the winding factor in the case of single layer windings is $K^{sl}_{wf7} = K^{sl}_{wf} \simeq 0.9659258$.

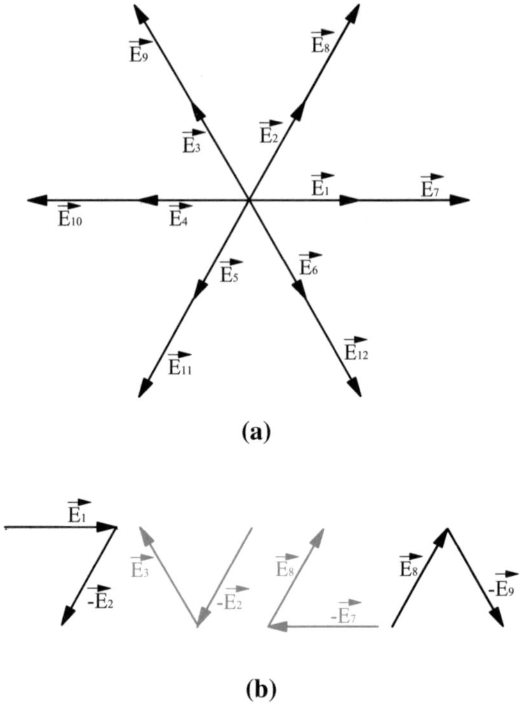

Fig. 3.14 Star of slots
based-prediction of the
winding factor of the
sub-harmonic of rank two of
the back-EMF induced in the
double-layer concentrated
winding characterized by
$N_s = 12$, $p = 5$, and $q = 3$

(a)

(b)

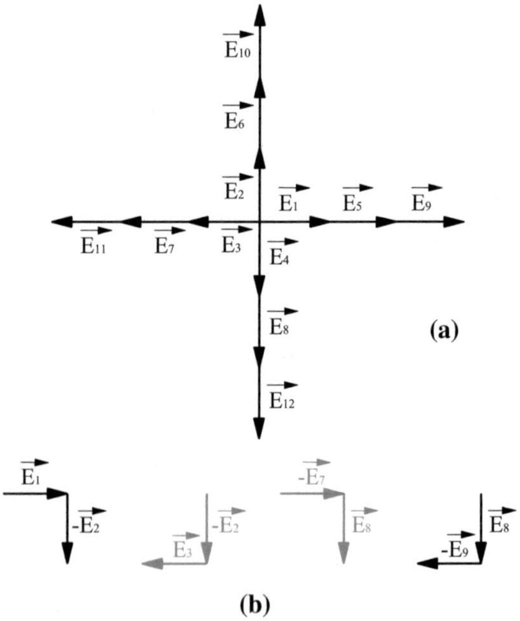

Fig. 3.15 Star of slots
based-prediction of the
winding factor of the
sub-harmonic of rank three
of the back-EMF induced in
the double-layer
concentrated winding
characterized by $N_s = 12$,
$p = 5$, and $q = 3$

(a)

(b)

Fig. 3.16 Star of slots based-prediction of the winding factor of the super-harmonic of rank seven of the back-EMF induced in the double-layer concentrated winding characterized by $N_s = 12$, $p = 5$, and $q = 3$

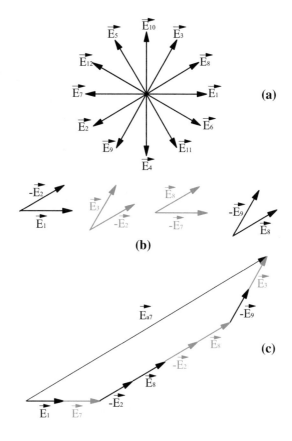

3.4 MMF Formulation and Analysis: A Case Study

Let us consider the FSPMM characterized by $N_s = 12$, $p = 5$, and $q = 3$. Following the determination of the double- and single-layer distributions yielded by the star of slots approach, the MMFs generated by the resulting concentrated windings are investigated in the present paragraph.

3.4.1 Double-Layer Topology

Let us remind the armature winding distributed in 12 double-layer slots, as illustrated in Fig. 3.17.

Giving the fact that each phase is made up of N coils, and accounting for the notations considered in Fig. 3.17, the MMF $\mathcal{F}_{a1}$ produced by the quarter of phase "a" ($a1^+/a1^-$) is expressed as follows:

Fig. 3.17 Armature winding
arrangement in double-layer
slots of a FSPMM
characterized by $N_s = 12$,
$p = 5$, and $q = 3$

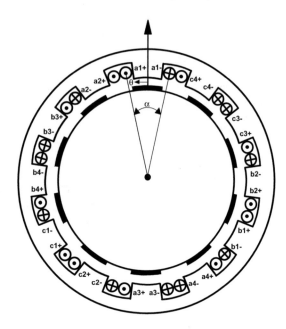

$$\mathcal{F}_{a1}(\theta) = \begin{cases} (2 - \beta)\,\dfrac{NI}{8} & -\dfrac{\alpha}{2} < \theta < \dfrac{\alpha}{2} \\ -\beta\,\dfrac{NI}{8} & \text{elsewhere} \end{cases} \tag{3.8}$$

where β is the coil pitch to the pole pitch ratio, such that:

$$\beta = \frac{\alpha}{\tau_{ps}} \tag{3.9}$$

Taking into account the hypothesis which considers that the magnetic circuit is
not saturated, one can apply the superposition theorem in order to obtain the MMF
$\mathcal{F}_a(\theta)$ of phase "a", as follows:

$$\mathcal{F}_a(\theta) = \mathcal{F}_{a1}(\theta) + \mathcal{F}_{a2}(\theta) + \mathcal{F}_{a3}(\theta) + \mathcal{F}_{a4}(\theta) \tag{3.10}$$

with:

$$\begin{cases} \mathcal{F}_{a2}(\theta) = -\mathcal{F}_{a1}(\theta - \frac{\pi}{6}) \\ \mathcal{F}_{a3}(\theta) = -\mathcal{F}_{a1}(\theta - \pi) \\ \mathcal{F}_{a4}(\theta) = \mathcal{F}_{a1}(\theta - \frac{7\pi}{6}) \end{cases} \tag{3.11}$$

It has been reported in [14] that the armature MMF $\mathcal{F}(\theta)$ is bipolar ($p_s = 1$).
Therefore, the armature pole pitch τ_{ps} is equal to π, leading to the following expression of the coal pitch to the pole pitch ratio β:

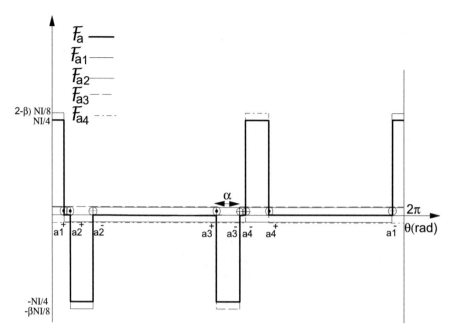

Fig. 3.18 Spatial repartitions of the MMFs produced by phase "a" coils fed by the maximum current I and their sum $\mathcal{F}_a(\theta)$ (bold line)

$$\beta = \frac{\alpha}{\pi} \tag{3.12}$$

Let us call γ the slot opening. Referring to Fig. 3.17, one can write:

$$\alpha + \frac{\gamma}{2} = \frac{\pi}{6} \tag{3.13}$$

which yields:

$$\gamma = \frac{\pi}{3} - 2\alpha = \pi(\frac{1}{3} - 2\beta) \tag{3.14}$$

Figure 3.18 shows the spatial repartition of the MMF induced by phase "a".

Now let us consider the spatial repartition of the MMFs produced by phases "b" and "c", which are deduced from the one produced by phase "a" as follows:

$$\begin{cases} \mathcal{F}_b(\theta) = -\frac{1}{2}\,\mathcal{F}_a(\theta + \frac{2\pi}{3}) \\ \mathcal{F}_c(\theta) = -\frac{1}{2}\,\mathcal{F}_a(\theta - \frac{2\pi}{3}) \end{cases} \tag{3.15}$$

Then, the superposition of $\mathcal{F}_a$, $\mathcal{F}_b$ and $\mathcal{F}_c$ has led to the spatial profile of the resulting armature MMF $\mathcal{F}(\theta)$ shown in Fig. 3.19.

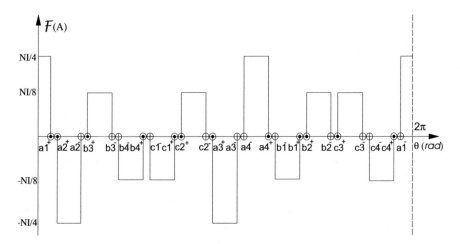

Fig. 3.19 Spatial repartition of the MMF created by a double layer FSPMM characterized by $N_s = 12$, $p = 5$, and $q = 3$, fed by sinusoidal currents

Let us consider the *Fourier* expansion of $\mathcal{F}_{a1}$:

$$\mathcal{F}_{a1}(\theta) = \frac{NI}{2\pi}\left\{\sin(\beta\tfrac{\pi}{2})\cos(\theta) + \tfrac{1}{2}\sin(\beta\pi)\cos(2\theta) + \ldots + \tfrac{1}{n}\sin(n\beta\tfrac{\pi}{2})\cos(n\theta)\right\} \quad (3.16)$$

Taking into account Eqs. (3.11), the *Fourier* expansions of $\mathcal{F}_{a2}$, $\mathcal{F}_{a3}$ and $\mathcal{F}_{a4}$ can be deduced from the one of $\mathcal{F}_{a1}$ as follows:

$$\begin{cases} \mathcal{F}_{a2}(\theta) = -\frac{NI}{2\pi}\left\{\sin(\beta\tfrac{\pi}{2})\cos(\theta - \tfrac{\pi}{6}) + \ldots + \tfrac{1}{n}\sin(n\beta\tfrac{\pi}{2})\cos(n(\theta - \tfrac{\pi}{6}))\right\} \\ \mathcal{F}_{a3}(\theta) = -\frac{NI}{2\pi}\left\{\sin(\beta\tfrac{\pi}{2})\cos(\theta - \pi) + \ldots + \tfrac{1}{n}\sin(n\beta\tfrac{\pi}{2})\cos(n(\theta - \pi))\right\} \quad (3.17) \\ \mathcal{F}_{a4}(\theta) = \frac{NI}{2\pi}\left\{\sin(\beta\tfrac{\pi}{2})\cos(\theta - \tfrac{7\pi}{6}) + \ldots + \tfrac{1}{n}\sin(n\beta\tfrac{\pi}{2})\cos(n(\theta - \tfrac{7\pi}{6}))\right\} \end{cases}$$

Accounting for Eqs. (3.16) and (3.17), one can write the general term $\mathcal{F}_{an}(\theta)$ of the *Fourier* expansion of $\mathcal{F}_a$:

$$\mathcal{F}_{an}(\theta) = \frac{NI}{2\pi}\frac{\sin(n\,\beta\,\tfrac{\pi}{2})}{n}\left\{\cos(n\theta) - \cos(n(\theta - \tfrac{\pi}{6})) - \cos(n(\theta - \pi)) + \cos(n(\theta - \tfrac{7\pi}{6}))\right\} \quad (3.18)$$

and then deduce those of phases "b" and "c", such that:

$$\begin{cases} \mathcal{F}_{bn}(\theta) = -\frac{NI}{4\pi}\frac{\sin(n\,\beta\,\tfrac{\pi}{2})}{n}\left\{\cos(n(\theta - \tfrac{2\pi}{3})) - \cos(n(\theta - \tfrac{5\pi}{6})) - \right. \\ \qquad\qquad\qquad\qquad\qquad\qquad \left. \cos(n(\theta - \tfrac{5\pi}{3})) + \cos(n(\theta - \tfrac{11\pi}{6}))\right\} \\ \mathcal{F}_{cn}(\theta) = -\frac{NI}{4\pi}\frac{\sin(n\,\beta\,\tfrac{\pi}{2})}{n}\left\{\cos(n(\theta - \tfrac{2\pi}{3})) - \cos(n(\theta - \tfrac{\pi}{2})) - \right. \\ \qquad\qquad\qquad\qquad\qquad\qquad \left. \cos(n(\theta - \tfrac{\pi}{3})) + \cos(n(\theta - \tfrac{\pi}{2}))\right\} \end{cases} \quad (3.19)$$

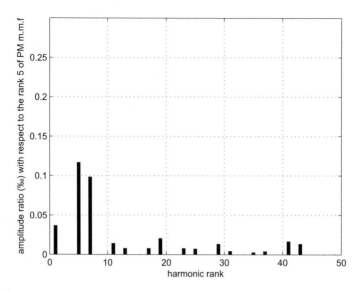

Fig. 3.20 Spectrum of the armature MMF $\mathcal{F}(\theta)$ of a double-layer FSPMM characterized by $N_s = 12$, $p = 5$, and $q = 3$, and fed by sinusoidal currents, for $\beta = \frac{1}{8}$

The superposition of $\mathcal{F}_{an}(\theta)$, $\mathcal{F}_{bn}(\theta)$ and $\mathcal{F}_{cn}(\theta)$ given by Eqs. (3.18) and (3.19) leads to the following general form of MMF created by the double-layer winding:

$$\mathcal{F}_n^{dl}(\theta) = A_n^{dl} \sin\left(n(\theta - \frac{\pi}{12})\right) \tag{3.20}$$

where:

$$A_n^{dl} = \frac{NI}{\pi} \left(\frac{1 + (-1)^{n+1}}{n}\right) \left(\cos\left(n\frac{2\pi}{3}\right) - 1\right) \sin\left(n\beta\frac{\pi}{2}\right) \sin\left(\frac{n\pi}{12}\right) \tag{3.21}$$

Accounting for Eq. (3.20), one can represent the spectrum of $\mathcal{F}(\theta)$ as illustrated in Fig. 3.20.

3.4.2 Single-Layer Topology

The transition from the double-layer distribution to the single-layer one is achieved using the approach developed in Sect. 3.2.4 which yields the winding distribution illustrated in Fig. 3.21.

Under sinusoidal current supply with a maximum current in phase "a", the armature generates the MMF spatial repartition shown in Fig. 3.22.

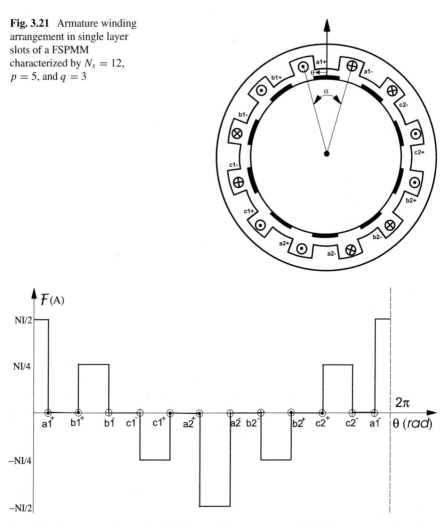

Fig. 3.21 Armature winding arrangement in single layer slots of a FSPMM characterized by $N_s = 12$, $p = 5$, and $q = 3$

Fig. 3.22 Spatial repartition of the MMF created by a single-layer FSPMM characterized by $N_s = 12$, $p = 5$, and $q = 3$, fed by sinusoidal currents

Giving the fact that:

$$\mathcal{F}_{a2}(\theta) = -\mathcal{F}_{a1}(\theta - \pi) \tag{3.22}$$

and accounting for Eq. (3.15), one can derive the general terms of the *Fourier* transform of phases "a", "b" and "c", as follows:

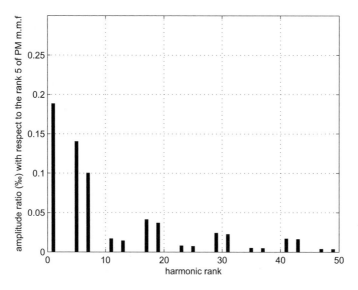

Fig. 3.23 Spectrum of the armature MMF $\mathcal{F}(\theta)$ of a single-layer FSPMM characterized by $N_s = 12$, $p = 5$, and $q = 3$, and fed by sinusoidal currents, for $\beta = \frac{1}{8}$

$$
\begin{cases}
\mathcal{F}_{an}(\theta) = \frac{4}{\pi} \frac{NI}{4} \frac{(1+(-1)^{n+1})}{n} \sin(n\beta\frac{\pi}{2})\cos(n\theta) \\[2mm]
\mathcal{F}_{bn}(\theta) = -\frac{1}{2}\frac{4}{\pi}\frac{NI}{4}\frac{(1+(-1)^{n+1})}{n}\sin(n\beta\frac{\pi}{2})\cos(n(\theta+\frac{2\pi}{3})) \\[2mm]
\mathcal{F}_{cn}(\theta) = -\frac{1}{2}\frac{4}{\pi}\frac{NI}{4}\frac{(1+(-1)^{n+1})}{n}\sin(n\beta\frac{\pi}{2})\cos(n(\theta-\frac{2\pi}{3}))
\end{cases} \quad (3.23)
$$

Finally, the general term of the *Fourier* expansion of the resulting MMF created by a single-layer winding is written as:

$$
\mathcal{F}_n^{sl}(\theta) = A_n^{sl}\cos(n\theta) = \frac{NI}{\pi}\left(\frac{1+(-1)^{n+1}}{n}\right)\left(1-\cos\left(n\frac{2\pi}{3}\right)\right)\sin(n\beta\frac{\pi}{2})\cos(n\theta) \quad (3.24)
$$

that gives the spectrum shown in Fig. 3.23.

3.4.3 Double- Versus Single-Layer Topologies

This paragraph is aimed at a comparison between the harmonic contents of both double- and single-layer topologies. To do so, let us consider the ratio $\left(\mathcal{R}_{sl}^{dl}\right)_n = \left|\frac{A_n^{dl}}{A_n^{sl}}\right|$ which is deduced from Eqs. (3.21) and (3.24), as:

Table 3.3 Evolution of ratio $\left(\mathcal{R}_{sl}^{dl}\right)_n$ for the fundamental and the harmonics lower than 20

	Fundamental	Harmonics					
	$n = 5$	$n = 1$	$n = 7$	$n = 11$	$n = 13$	$n = 17$	$n = 19$
$\left\lvert\frac{A_n^{dl}}{A_n^{sl}}\right\rvert$	0,9659	0,5366	0,9659	0,2588	0,2588	0,9659	0,9659

$$\left\lvert\frac{A_n^{dl}}{A_n^{sl}}\right\rvert = \left\lvert\frac{\frac{NI}{\pi}\left(\frac{1+(-1)^{n+1}}{n}\right)\left(\cos\left(n\frac{2\pi}{3}\right)-1\right)\sin\left(n\beta\frac{\pi}{2}\right)\sin\left(\frac{n\pi}{12}\right)}{\frac{NI}{\pi}\left(\frac{1+(-1)^{n+1}}{n}\right)\left(1-\cos\left(n\frac{2\pi}{3}\right)\right)\sin(n\beta\frac{\pi}{2})}\right\rvert = \left\lvert\sin\left(\frac{n\pi}{12}\right)\right\rvert \quad (3.25)$$

which highlights that the amplitudes of all harmonics of the *Fourier* expansion of the double-layer topology MMF are lower than the ones of the MMF harmonics yielded by the single-layer topology. This inevitably influences the torque production capability with a reduction of the mean torque allied to a decrease of the fundamental amplitude as well as a reduction of torque ripple allied to a decrease of the amplitudes of the harmonics. Table 3.3 gives an accurate idea about the impact of substituting a single-layer arrangement by a double-layer one, on the harmonic content of the air gap MMF created by the armature.

3.5 Conclusion

This chapter was devoted to the arrangement of the armature winding of FSPMMs, which has been achieved considering the star of slots approach. Following the winding arrangement, the star of slots was used for the investigation of the back-EMF harmonic content as well as for the determination of the winding factors of the fundamental and harmonic backs-EMFs. Finally a case study was treated with emphasis on the spatial repartition and harmonic content of the MMF produced by the armature arranged in double- as well as in single-layer slots. It has been found that the amplitudes of all harmonics of the *Fourier* expansion of the double-layer topology MMF are lower than the ones of the MMF harmonics yielded by the single-layer topology, with:

- a decrease of the fundamental amplitude not exceeding 3.41%,
- a decrease of the amplitude of the subharmonic of rank 1 reaching up 46.34%,
- a decrease of the amplitudes of the harmonics of ranks 11 and 13 reaching up 74.12%.

These results clearly highlight that while the mean value of the electromagnetic torque developed by the double-layer topology is slightly lower than the one produced by the single-layer topology, the torque ripple of the latter is higher.

References

1. El-Refaie AM, Jahns TM (2004) Optimal flux weakening in surface PM machines using concentrated windings. In: Proceedings of the IEEE industry applications society annual meeting. Seattle, USA
2. El-Refaie AM, Jahns TM, McCleer PJ, McKeever JW (2006) Experimental verification of optimal flux weakening in surface PM machines using concentrated windings. IEEE Trans Ind Appl 21(2):362–369
3. Patent no. 92958, "Mechrphasenmaschine mit ungleicher ankerspulen und polzadl," *Deutsches Reichspatent*, 1895
4. Cros J, Viarouge P (2002) Synthesis of high performance PM motors with concentrated windings. IEEE Trans Energy Convers 17:248–253
5. EL-Refaie AM, Shah MR, Qu R, Kern JM (September 2007) Effect of number of phases on losses in conducting sleeves of high speed surface PM machine rotors. In: Proceedings of the IEEE industry applications society annual meeting. New Orleans, USA, pp. 1522–1529
6. Libert F, Soulard J (September 2004) Investigation on pole-slot combinations for permanent magnet machines with concentrated windings. In: Proceedings of the international conference on electrical machines (ICEM'04). Cracow, Poland
7. Bianchi N, Dai Pre M (2006) Use of the star of slots in designing fractional-slot single-layer synchronous motors. IEE Proc Electr Power Appl 153(3): 997–1006
8. Bianchi N, Bolognani S, Grezzani G (2006) Design considerations for fractional-slot winding configurations of synchronous machines. IEEE Trans Ind Appl 42(4):997–1006
9. Richter R (1952) Lehrbuch der wicklungen elektrischer machinen. W. Bucherei Edition, Karlsruhe
10. Liwschitz-Garik M, Whipple CC (1960) Electric machinery, A-C Machines, vol. II. D. Van Nostrand Company Inc. New York
11. Bianchi N, Dai Pre M, Alberti L, Fornasiero E (September 2007) Theory and design of fractional-slot PM machines. In: Industry applications society annual meeting. New Orleans, USA (Tutorial course notes)
12. Magnussen F, Sadarangani C (June 2003) Winding factors and Joule losses of permanent magnet machines with concentrated windings. In: Proceedings of the IEEE international electric machines and drives conference, vol. 1. Madison, Winsconsin, USA, pp. 333–339
13. Zhu ZQ (March 2009) Fractional slot Permanent magnet brushless machines and drives for electric and hybrid propulsion systems, Plenary Session. In: Proceedings of the fourth international conference and exhibition on ecological vehicles and renewable energies. Monte-Carlo, Monaco
14. Ben Hamadou G, Masmoudi A, Abdennadher I, Masmoudi A (2009) Design of a single-stator dual-rotor permanent magnet machine. IEEE Trans Magn 45(1):127–132
15. Abdennadher I, Masmoudi A (2015) Armature design of low voltage FSPMSMs: an attempt to enhance the open-circuit fault tolerance capabilities. IEEE Trans Ind Appl 51(6):4392–4403

Printed in the United States
By Bookmasters